The Government of Natural Resources

The Nature | History | Society series is devoted to the publication of high-quality scholarship in environmental history and allied fields. Its broad compass is signalled by its title: *nature* because it takes the natural world seriously; *history* because it aims to foster work that has temporal depth; and *society* because its essential concern is with the interface between nature and society, broadly conceived. The series is avowedly interdisciplinary and is open to the work of anthropologists, ecologists, historians, geographers, literary scholars, political scientists, sociologists, and others whose interests resonate with its mandate. It offers a timely outlet for lively, innovative, and well-written work on the interaction of people and nature through time in North America.

General Editor: Graeme Wynn, University of British Columbia

A list of titles in the series appears at the end of the book.

The Government of Natural Resources

Science, Territory, and State Power in Quebec, 1867–1939

STÉPHANE CASTONGUAY

TRANSLATED BY KÄTHE ROTH

FOREWORD BY GRAEME WYNN

UBC Press • Vancouver • Toronto

© UBC Press 2021

Originally published as *Le gouvernement des ressources naturelles: sciences et territorialités de l'État québécois 1867–1939* © 2016 Presses de l'Université Laval

All rights reserved. No part of this publication may be reproduced, stored in a retrieval system, or transmitted, in any form or by any means, without prior written permission of the publisher, or, in Canada, in the case of photocopying or other reprographic copying, a licence from Access Copyright, www.accesscopyright.ca.

30 29 28 27 26 25 24 23 22 21   5 4 3 2 1

Printed in Canada on FSC-certified ancient-forest-free paper (100% post-consumer recycled) that is processed chlorine- and acid-free.

Library and Archives Canada Cataloguing in Publication

Title: The government of natural resources : science, territory, and state power in Quebec, 1867–1939 / Stéphane Castonguay ; translated by Käthe Roth.
Other titles: Gouvernement des ressources naturelles. English
Names: Castonguay, Stéphane, author. | Roth, Käthe, translator.
Series: Nature, history, society.
Description: Series statement: Nature, history, society | Translation of: Le gouvernement des ressources naturelles : sciences et territorialités de l'État québécois 1867–1939. | Includes bibliographical references and index.
Identifiers: Canadiana (print) 2021013836X | Canadiana (ebook) 20210138424 | ISBN 9780774866309 (hardcover) | ISBN 9780774866316 (softcover) | ISBN 9780774866323 (PDF) | ISBN 9780774866330 (EPUB)
Subjects: LCSH: Earth sciences – Government policy – Québec (Province) – 19th century. | LCSH: Earth sciences – Government policy – Québec (Province) – 20th century. | LCSH: Administrative agencies – Québec (Province) – History – 19th century. | LCSH: Administrative agencies – Québec (Province) – History – 20th century. | LCSH: Natural resources – Québec (Province) – Management – History – 19th century. | LCSH: Natural resources – Québec (Province) – Management – History – 20th century.
Classification: LCC QE193 .C3713 2022 | DDC 557.1409/034—dc23

Canadä

UBC Press gratefully acknowledges the financial support for our publishing program of the Government of Canada (through the Canada Book Fund), the Canada Council for the Arts, and the British Columbia Arts Council.

This book has been published with the help of a grant from the Canadian Federation for the Humanities and Social Sciences, through the Awards to Scholarly Publications Program, using funds provided by the Social Sciences and Humanities Research Council of Canada. We acknowledge the financial support of the Government of Canada through the National Translation Program for Book Publishing, an initiative of the Roadmap for Canada's Official Languages 2013–2018: Education, Immigration, Communities, for our translation activities.

UBC Press
The University of British Columbia
2029 West Mall
Vancouver, BC V6T 1Z2
www.ubcpress.ca

This book is dedicated
to the memory of Andréanne Deslippe
(1993–2020).

Contents

List of Figures and Tables / ix

Foreword / xi
Graeme Wynn

Acknowledgments / xxiii

Introduction / 3

1 The Administrative Capacities of the Quebec State: Specialized Personnel and Technoscientific Interventions / 20

2 The Invention of a Mining Space: Geological Exploration and Mineralogical Knowledge / 43

3 Soil Classification and Separation of Forest and Colonization Areas: Scientific Forestry and Reforestation / 71

4 Surveillance and Improvement of Fish and Game Territories: Conservation of Wildlife Resources / 95

5 Regionalization and Specialization of Agricultural Production: Disseminating Agronomic Knowledge / 124

Conclusion: Knowledge, Power, and Territory / 156

Appendix: Identification of Technoscientific Activities in the Public Accounts (1896–1940) / 159

Notes / 161

Bibliography / 188

Index / 200

Figures and Tables

Figures

1.1 The provincial fish hatchery at Saint-Faustin / 28
1.2 Technoscientific interventions and exploitation of agricultural resources / 32
1.3 Technoscientific interventions and exploitation of forestry, mining, and wildlife resources / 33
1.4 Field laboratory in Granby / 35
1.5 Berthierville nursery: rain gauge / 38
1.6 Laboratory (Department of Mines) / 38
2.1 George Dawson's unexplored territories / 47
2.2 Mining as a national industry / 51
2.3 Path of the National Transcontinental Railway and Grand Trunk Pacific Railway / 55
2.4 Members of the Chibougamau Mining Commission at work / 59
2.5 The creation of northwestern Quebec as a mining space / 64
2.6 Sites of explorations by the Quebec Bureau of Mines and institutional affiliations of geologists / 65
2.7 Mineral map of Quebec, 1914 / 67
2.8 Mineral map of Quebec, 1930 / 68
3.1 Map of forest reserves / 77
3.2 Berthierville nursery: students depart to conduct inventory and study the nearby bogs / 82
3.3 Number and area of township forest reserves in Quebec, 1911–39 / 90
3.4 Distribution of township forest reserves in Quebec, 1911–39 / 90

3.5 Nursery in the Parke Reserve / 92
3.6 Pile of pulpwood in the Parke Reserve / 93
4.1 Quebec public revenues from fish and game club activities / 97
4.2 Number of wardens hired by the Fisheries and Game Service / 104
4.3 Number of wardens employed by the Fisheries and Game Service and by private clubs / 105
4.4 Amounts of fines collected / 105
4.5 Distribution of game wardens and fish wardens by district / 107
4.6 Division and distribution of Fisheries and Game Service personnel assigned to surveillance by county, 1922 and 1932 / 108
4.7 Enclavement of Laurentide National Park by private clubs / 113
4.8 B.W. Taylor stocking Lake Sorcier with ouananiche fry / 118
4.9 Distribution of ouananiche eggs and fry in Quebec, 1931–35 / 119
4.10 Distribution of salmon fry in Quebec, 1929–33 and 1934–38 / 120
4.11 The provincial fish hatchery at Saint-Faustin in the Laurentians / 121
5.1 Number of official agronomists, 1913–32 / 130
5.2 Map of the province of Quebec showing agronomic districts, 1917–18 / 131
5.3 Map of the twenty-three agricultural districts in 1929 / 135
5.4 Map of the twenty agronomic districts in 1933 and location of residence of the regional agronomist for each district / 137
5.5 Agronomic Service's activities, annual averages per agronomist, 1920–32 / 139
5.6 Apple production per county, 1921 and 1931, and demonstration sites, 1918 and 1928 / 145
5.7 Protection of orchards in Rougemont: spraying with sulfur fungicide / 147
5.8 Contribution by district to provincial poultry production, 1901–41 / 152
5.9 Poultry stations in operation, by county, 1921 and 1931 / 153

Tables

1.1 Technoscientific personnel in the internal service / 22
1.2 Technoscientific personnel in the external service / 24
1.3 Technoscientific interventions / 31
3.1 Distribution of seedlings at the Berthierville nursery / 86

FOREWORD
Science in Action
Graeme Wynn

CANADA WAS *"Made Modern."* According to an important recent collection of essays edited by historians of science Edward Jones-Imhotep and Tina Adcock, this means two things.[1] First: the country brought into being by the British North America Act of 1867 was shaped by a specific set of late-nineteenth- and early-twentieth-century circumstances. These included the rise of the nation-state as a territorial and social entity; a growing conviction that the world could be improved by human intervention; the transformation of subjects into citizens; the development of new ways of administering society and space; a reorientation from the past to the future; the acceptance of the market as an arbiter of value; the embrace of rationality as an organizing principle for human affairs; and a broad liberal commitment to individualism, personal freedom, and formal equality. Together, these traits constituted a "will to modernize" – an injunction to use the human and non-human resources of the nation to realize its potential.[2]

Second: modernization, as a process driven by the spirit of improvement, is ongoing, a goal to be pursued, rather than a destination to be achieved. As such, it implies continuing engagement. Even places that have "always been modern," as signified above, have work to do if they aspire to remain modern through time. In a country as large and as diverse as Canada, the challenges and results of this work were, and are, deeply uneven, especially as the country that came into being in 1867 was no *tabula rasa*. Indigenous people had occupied much of this northern territory for millennia before the mid-nineteenth century. Their ways of life, their traditions, their spiritual and intellectual beliefs, their connections with the environments in which they lived were, generally, radically different from those of the newcomers, and the worlds of which those people were a part.[3]

Nor were "newcomers" of a piece. Many claimed several generations of familial connection to the land they occupied in 1867. Others counted their time in the new confederation in weeks or months rather than years. Many things differentiated them, beyond the length of their exposure to Canadian air. Some recent arrivals were refugees from deeply entrenched, fundamentally conservative, but crumbling and threatened settlements on the agricultural margins of the United Kingdom; some were fleeing urban poverty or the threat of immiseration as industrialization undermined the bases of their livelihoods; yet others – through birth, education, occupation, curiosity, or chance – were fully cognisant of the intellectual, commercial, and political currents quickening the mid-nineteenth-century Atlantic world.[4] In some Canadian settings, settlers clung to remnants of old-world traditions; in others, immigrants seemed to abjure their pasts with abandon.

Modernity was complex, shape-shifting, evolving; with change at its heart, it could hardly be otherwise. Changes precipitated by the commitment to progress had radically different impacts, across space and through time; they were variously embraced and/or resisted; and they left markedly different, "often inequitable imprints on gendered, racialized, classed, aged and aging, and regionalized bodies."[5] Study after detailed study of early Canadian development has shown (as contributors to *Made Modern* recognize) that "learning to be modern could be difficult, stressful, and even frightening," and that some resisted the threat it presented by adopting behaviours that later historians have characterized as "anti-modernist."

Stéphane Castonguay's examination of the ways in which science, territory, and state power became entangled in the administration of Quebec's natural resources before the Second World War both narrows and broadens the wide-ranging conversation about science, technology, and modernity broached in *Made Modern*. Although Castonguay's book is replete with comparative allusions to other provinces of Canada, its focus is resolutely on Quebec. It is also, and equally resolutely, concerned with the exploitation of that province's "natural resources" – its minerals and trees, its fish and game, and its agricultural potential.[6] Both books engage repeatedly with science, technology, and the environment, but they tend to do so in different ways, using different vocabularies and working at different scales. These differences are due, in part, to the different provenances of the two books, one the work of a single author, the other a collection of discrete contributions. But there is more to it than this. Published originally in French, in 2016, *The Government of Natural Resources* bears the identifiable traces, even in translation, of a distinctive style of discourse; it draws on ideas less familiar to anglophone than to francophone scholarship; and its arguments turn on concepts that warrant brief contextualization here. Foremost among these, perhaps, is the widely, but variously, used term "technoscience."

Seeking a novel and intelligible framework for understanding historical patterns common to science, technology, and medicine in the West since the Renaissance, the English historian of medicine John V. Pickstone focused on their ways of knowing the world: natural history as description and classification; analysis as the physical or intellectual taking apart of objects or systems to identify their elements; and experimentalism as the rearrangement or recombination of elements to produce new entities.[7] Although Pickstone saw natural history, analysis, and experimentalism as broadly sequential ways of knowing, he was at pains to point out that, over time, each new, dominant form *dis*placed (reduced the importance of) rather than *re*placed its predecessor; science, technology, and medicine are typically plural enterprises.

Alongside this tripartite classification, Pickstone placed "technoscience." Although he drew upon Bruno Latour's formulation in his 1987 book *Science in Action,* his use of the term was less radical than Latour's. Both scholars take technoscience to mark the intricate entanglements of science, technology, and society, but the latter synthesizes ideas developed by the philosophers Gaston Bachelard in France and Gilbert Hottois in Belgium with a raft of contemporary work in science and technology studies to dissolve the distinction between people and things, challenge the divide between nature and culture, and deem humans, microbes, and machines "actants" equally capable of exercising power and shaping scientific networks. Pickstone generally used "technoscience" in a more restricted sense, to emphasize the synergetic collaborations of government, academic (scientific), and commercial interests.

Intent on making his point about the potential significance of such collaborations, Pickstone located the apogee of technoscience in the "dense intertwining of universities, industry and government" in massive organizations or projects, and exemplified these with the CERN high-energy particle accelerator in Geneva or, more broadly, by reference to the "military-industrial" and "medico-industrial" complexes that have dominated the production of scientific commodities (as well as academic and industrial worlds) through most of the last seventy-five years. But Pickstone the historian was quick to acknowledge that such collaborations have antecedents. Early voyages of exploration sponsored by monarchs, commercial adventurers, or the state (think of Queen Isabella and Columbus, the Dutch East India Company, or the voyage of the *Beagle*) were technoscientific enterprises with inventorial purposes. More concertedly, academic, government, commercial, scientific, and technical interests came together in nineteenth-century Britain to address, through analysis and experimentation, difficult challenges of steamship building and transoceanic telegraph communication.[8] Until the middle years of the nineteenth century, however, technoscientific networks "were still very marginal to government, to the direction of work in most universities, and to most industrial companies."[9]

Collaborations between science and technology have certainly been integral to the development of Canada. The first fishermen, explorers, traders, and colonizers to reach the northeastern foreland of the American continent depended upon the (then) sophisticated technology, refined through the ages, of the sailing vessels that carried them across the Atlantic, and the nascent science of celestial navigation. As they entered the territory, they borrowed Indigenous technologies (most famously the snowshoe and the canoe), drew knowledge from Indigenous informants, and used early scientific instruments such as the astrolabe and the quadrant to inscribe their developing, increasingly systematized knowledge on maps and charts. By the latter decades of the eighteenth century, new technologies (the chronometer, the sextant) allowed more accurate measurement of the world. Whether their knowledge was represented in maps, books, sketches, collections, or the accumulation of a more intricate awareness of local surroundings, people knew the territory that would become Canada through this extended period as natural historians. They sought to inventory, describe, and classify what they had. Yet insofar as state and commercial interests were involved in this project (think of the support France offered explorers and seigneurs or of the activities of the Hudson's Bay Company), those inclined to broad conceptualizations might discern nascent forms of technoscience at work.[10]

Pickstone's ways of knowing also help to illuminate Canadian technoscientific history. The natural history impulse persisted well into the nineteenth century. We find it expressed in *The Canadian Naturalist: A Series of Conversations on the Natural History of Lower Canada*, a book published in 1840 by Henry Gosse, based on his time as a farmer in Compton, Canada East, where he was known as "that crazy Englishman who goes about picking up bugs."[11] It drove the specimen-collecting enthusiasms and countryside excursions of countless members of natural history societies that flourished across the country in the second half of the century; indeed, the Natural History Society of Prince Edward Island, formed to awaken "interest in the study of natural objects," came into being as late as 1889.[12] When the Geological Survey of the United Provinces of Canada was founded in 1842, its mandate was to furnish "a full and scientific description of the country's rocks, soils, and minerals, to prepare maps, diagrams, and drawings, and to collect specimens to illustrate the occurrences."[13] This remained an important commitment of the Geological Survey of Canada long after Confederation, as its employees fanned out to "read the rocks" across the new nation's territory.[14]

Even in the 1840s, however, the first director of the Geological Survey, William Logan, deployed the analytical methods he had developed in Wales to produce accurate cross-sections of strata beneath the surface and to infer the ages of rocks.[15] Canadians also began to alter the hydrology of rivers and streams, experimenting in hope of increasing the runs of salmon in dry seasons; a couple of decades later, Canadian fish hatcheries began to gather and fertilize fish eggs, using techniques developed in Germany and refined in France in the eighteenth and nineteenth centuries, to restore stocks and introduce fish to new lakes and rivers.[16] In the 1860s, Samuel Wilmot, a farmer, merchant, and local government official in Newcastle Township, Canada West, built several wooden troughs in the basement of his farmhouse and piped in water from nearby Wilmot Creek to hatch the spawn of four salmon. With Confederation, the federal government began to support his efforts, and in 1868 he became a fishery overseer in the Department of Marine and Fisheries, with special responsibility for his hatchery.[17] Within ten years, Wilmot claimed that his expanded and elaborate labyrinth of ponds and raceways – intended to boost the commercial salmon fishery – was "the most complete and systematically arranged fish-breeding establishment on the continent."[18]

This was no small thing, and in its combination of science, government support, and commercial orientation, it offered a portent of things to come. Defining technoscience as "the interlacing of science and technique, both in artifacts and in the networking of actors" (Introduction, note 9), Castonguay focuses his inquiries, in the pages that follow, on this period of emergent "inventive, intense and self-perpetuating synergies" among academic, industrial, and state interests. Although his work certainly considers the roles of science and government in shaping nature, and the ways in which the state and its agents defined, inventoried, and extracted natural resources, he eschews extended engagement with Latour's concerns about the dissolution of the nature-culture divide, hybridity, and so on.[19] His interest – as he has written elsewhere – is on that period when "government leaders wanted to enlarge and diversify the state apparatus to intervene more directly in the nation's economic affairs," and scientists convinced decision-makers that their research "was the key to national prosperity by solving problems related to industrial production and resource conservation."[20]

In Canada (as in much of the rest of the developed world), these commitments flourished between 1870 and 1939, years marked by the rapid advance and growing institutionalization of science, heated debates about the nature and purpose of education, and concerted efforts to extend and consolidate the administrative powers of the state. As Britain and other European countries established new universities and polytechnics, and created departments and degrees for the advancement of scientific knowledge and training, some Canadians began to question the practicality of prevailing, traditional, liberal arts models of education rooted in religion and metaphysics.

Setting the terms of a discussion that would transform universities and Canadian education in the decades to come, in 1870 J.W. Dawson, principal of McGill University, described the move from "mere apprenticeship" in the industrial arts to advanced intellectual training in science as "the greatest educational movement of our time." Mathematicians, physicists, chemists, botanist, zoologists, and geologists were appointed piecemeal in universities across the country in the decades that followed. Many began in departments of Natural Science; the first chemists at McGill University were in the medical school, and Chemistry only attained departmental status in that institution in 1908. Meanwhile, provincial school systems wrestled with the challenges of training students in agriculture, mechanics, and manufacturing. On the eve of the First World War, a Royal Commission outlined a blueprint for education in schools combining "intellectual development with preparation for entrance into industrial society," but policy-makers fumbled the implementation of thoroughly modern curricula for technical education.[21]

In a country still heavily dependent upon the extraction of natural resources, instrumental views of science prevailed. It was valuable insofar as it facilitated economic growth. To this end, the University of Toronto appointed a professor of agriculture in 1852 – but his tenure was short-lived as practical farmers resisted book learning. A few years later, the École d'agriculture de Sainte-Anne-de-la-Pocatière opened its classrooms and model farm in the district of Kamouraska; some 573 students enrolled between 1859 and 1912, but the college awarded only 16 diplomas and 27 certificates of aptitude.[22] By the third quarter of the nineteenth century, there was more inclination to accept science in action as a spur to productivity. The Ontario School of Agriculture was established in 1874 and was followed by similar establishments in Nova Scotia (Truro, in 1885) and Quebec (Oka, in 1893). By 1890, there were five Dominion Experimental Farms across the country mandated to conduct "experiments or researches as might benefit agriculture."[23] In a similar vein, and after much wrangling, a School of Practical Science was affiliated with University College in the University of Toronto in 1878. As the so-called Second Industrial Revolution gathered momentum in Canada from the 1880s, science was increasingly valued for its practical applications. Driving home the point, in 1894, the president of the Royal Society of Canada, the geologist G.M. Dawson, observed that the country was still "perhaps too young to afford public support to purely abstract researches in such subjects as chemistry, physics or biology, however valuable their possible results may be to the general knowledge of the world."[24]

As Suzanne Zeller has noted, however, these end-of-the-century years saw "engineers applying hydroelectricity's marvellous powers to everyday tasks; chemists refining, even synthesizing nature through pulp and paper, aniline dye, and textile industries; urban planners meeting unprecedented biochemical demands for sanitation systems; and a host of other science-based miracles."[25] Theoretical knowledge was now valuable currency. Technical know-how, acquired by observation and practice in any field, was increasingly exposed as inadequate by the "new light which scientific research is constantly throwing upon the subject."[26] Wealthy industrialists funded university laboratories and professorships, existing programs were upgraded, and new ones (such as the School of Mining and Agriculture at Queen's University, opened in 1893) were established. As the twentieth century wore on, science, "in its new guise as industrial research ... intertwined the interests of governments, educators, business, and labour."[27] Indeed, a 1949 history of chemistry in Canada gives most of its 500 pages to industrial chemistry and includes a chapter on chemistry and public services, "an account of the growth and nature of the National Research Council Laboratories, Experimental Farms, Forest Products Laboratories, other Federal laboratories [and] Provincial Research Councils."[28]

Although its title suggests that Stéphane Castonguay's book is more centrally concerned with efforts by the government of Quebec (the provincial state) to administer and regulate the exploitation of natural resources within its bounds than it is with the interconnections among university researchers, corporate interests, and society, a couple of caveats are in order. First, it is clear that the story told here, of science in the service of the state, is not a simple tale of instrumentalization. Science as an institution retained a form of agency and produced its own conditions of possibility within the state, especially as experimentalism displaced analysis as the dominant way of knowing in Canadian science and technology from the 1890s. And, second, although the pages that follow focus quite consistently on programs and policies formulated and implemented by civil servants (and beyond them, politicians) charged with responsibility for the development, extraction, or stewardship of the provincial estate, they also demonstrate that modernization of the state's capacity to administer its resources depended on the development, within government departments, of scientific expertise previously centred in the universities – and that by the early decades of the twentieth century, government scientists exercised considerable influence over institutions of higher education.

By the twentieth century, Castonguay's "technoscientific agents" played dramatically increased roles in constructing both the state apparatus and the effective territory of Quebec. This book is thus centrally concerned with issues of state formation and the organization and administration of space for different purposes; it is about governmentality and territoriality. These are large and well-worn topics, shaded with variant meanings. Some authors approach state formation as a process and document it by tracing the development of that "constellation of agencies and offices" created to articulate sovereign power. Others see the state as an agency of moral regulation and emphasize its role in the progressive, pervasive, and effective extension of social as well as political authority.[29] This latter formulation, associated with neo-Marxist sociology and sometimes described as a Durkheimian-Marxist perspective, comes closest to the idea of "governmentality" formulated by Michel Foucault and substantially adopted by Castonguay: here we have an understanding of power that goes beyond the top-down extension of state authority to include forms of social regulation exercised by disciplinary institutions and forms of knowledge.

In dealing with territoriality, Castonguay draws upon the ideas of two contrasting scholars, one American and one Swiss. From the University of Wisconsin geographer Robert Sack, who considered territoriality as a spatial strategy and defined it as the assertion and enforcement of control over a particular geographical area, Castonguay develops his approach to understanding the ways in which people instantiate their claims to resources by defining the spaces over which they have jurisdiction. From the Paris-born, University of Geneva–based geographer and theorist Claude Raffestin, who saw territoriality as "a set of relationships rooted in ties to the material environment and other people or groups, and mediated by existing techniques and representations," Castonguay draws his interest in exploring how scientists constructed territorialities for the identification and extraction of natural resources and disciplined people by defining codes of conduct for land settlement and resource exploitation.[30]

This is an unusual combination. Few scholars in the English-speaking world have paid much attention to Raffestin's work, and it is probably less known in Europe among scholars working in French than it is to those whose language is Italian.[31] Although both Sack and Raffestin made territoriality/*territorialité* a central focus of their inquiries, there are significant differences in their deployment of the term. Indeed, the political geographer Alexander Murphy has warned against seeing their positions as different views of the same phenomenon. Writing out of a discipline, Anglo-American geography, obsessed "with documenting and modelling locations, distances, flows, and networks," Sack was at pains to argue that "formalized or institutionalized spatial arrangements and partitions matter as well ... [and that] geography needed to go beyond treating them as objects of analysis and explain why they come into being and how they influence what happens." Accepting the modern conception of territories as discrete, bounded, identifiable spaces, Sack recognized them as deliberate human creations intended "to achieve certain social ends," and believed that the process of their creation – territoriality – offered an important window on circumstances important to the shaping of human societies.[32]

Raffestin, an independent-minded individual who draw inspiration from an eclectic range of scholars, held a rather broader and looser view of *territorialité* than Sack's analytic-behavioural conception of territoriality.[33] Influenced to some degree by the work of Jean Gottman, Michel Foucault, and Henri Lefebvre, as well as several continental philosophers and social theorists, Raffestin adopted a relational, post-positivist position that defies easy classification. He argues that "space becomes territory when it emerges out of social interaction," and insists that studies of territory are properly concerned with any "geographically-organized human activity."[34] In Alexander Murphy's thoughtful assessment, Sack is interested in "territorial divisions – actual or aspirational – and the processes that produce them," whereas Raffestin focuses on "the context within which social actions are embedded – a context that necessarily shapes territorial outcomes."[35] Given Raffestin's aversion to the objectivist and essentializing tendencies of spatial analysis in which Sack's work was rooted, their approaches are not easily reconciled.

Yet as Murphy acknowledges, the ideas of these two theorists overlap and there are potential synergies to be found in their shared interests in "power, iconography, and social relations." Castonguay works this ground, bringing his perspective as a historian of science to bear on Raffestin's and Sack's efforts to investigate how state and territory are mutually constructed. He begins with the observation (made by Thongchai Winichakul in his doctoral dissertation on the mapping of Siam) that maps often serve as models for, rather than models of, what they purport to represent.[36] Seen in this light, maps do more than represent space: they allow territories to be imagined in particular ways and facilitate acceptance of that vision. Cartographers make maps, but field scientists from several different disciplines also survey territories, to assess their productivity (by describing the human and natural resources they contain), and to develop plans for their rational use. Once scientists have enumerated the qualities of a territory, its "soil fertility, geological formations, ore deposits, or plant or animal species," Castonguay writes, "the state can define spaces, either to put them to a particular use or to protect resources for future exploitation or for conservation, often for the privilege of an exclusive group of users."

This was a multi-faceted enterprise. Fieldworkers coursed through spaces of interest to the state. Scientists and technicians with training in agronomy, botany, ecology, geology, fisheries, hydrology, forestry, pedology, zoology, and so on fanned out to evaluate assets; surveyors and engineers assisted their work, laying out roads, identifying sites for strategic construction, and defining limits and borders. The individual presence of each of these workers in any one place might be relatively fleeting, but as representatives of the state, they helped to legitimize its territorial interests. Other civil servants directed their attention to human populations, enumerating them, placing them on public lands, and legitimating their claims to particular parts of the whole. Another group of state officials – wardens with various warrants for surveillance and protection – travelled far and wide through newly valued terrain, making evident, on a local and daily basis, the authority of the state against those who would ignore or push back against its presence. Elsewhere, in government offices, fish hatcheries, experimental farms, demonstration sites, and laboratories, other scientists and technicians systematized, codified, and added to information from the field, before conveying what they had learned to settlers and workers on the resource frontiers, thus further impressing the hand of the state on affairs in its territory. Broadly, those who worked in offices and laboratories sought to acquire and codify information while those engaged in the field generally had more immediate, material concerns as they hatched fish, stocked rivers, cultivated nurseries, planted trees, and tested plant varieties – but the concerns of the two groups intersected and fed back one upon another. Beyond all of this, politicians and bureaucrats formulated, drafted, and ultimately oversaw the implementation of policies and regulations governing state resources. These were the technoscientific agents – the disparate group whose deployment and activities shaped construction of the Quebec state and its territory – around whom Castonguay builds the argument of this book.

Starting with a foray into the public accounts of Quebec, Castonguay offers an outline of the relative importance of the government's technoscientific expenditures between 1896 and 1940. The numbers of "agents" in each of the major resource sectors were never large, but they rose from barely 100 at the turn of the century to almost 750 in 1939. Generally there were more employees beyond Quebec City than within the corridors of government; agriculture and forestry were the dominant employees; and the mix of expenditures (for educational, descriptive, and experimental purposes) differed considerably over time within and among each of the resource sectors. Given the vastness of Quebec's natural resource estate, these few technoscientific agents achieved remarkable things. To take but one example, sixteen years after legislation initiating such activity, 60 percent of the province's leased forest had been inventoried, and there were silviculture plans for almost three-quarters of this area.

Such insights are important because they point to the larger truth, central to Castonguay's analysis, that the effectiveness of Quebec's technoscientific agents turned less upon direct policing and overt control than it did upon the establishment of "soft powers" that allowed people to make use of territories and resources within norms or regulatory frameworks built, increasingly, upon common understandings and shared interests. By tracing, in sequence through four chapters, the twinned histories of resource exploitation and governance in the province's agricultural, forestry, mining, and wildlife sectors, Castonguay shows how evolving terms of territorial occupation and exploitation supported the spatial and political expansion of the Quebec state, and how these terms followed a similar dynamic across the four sectors. "Far from grappling with a vast, undifferentiated territory," he writes in conclusion, Quebec's public administration identified and organized "a series of spaces" that its "technoscientific personnel were constantly reconstructing," in accordance with their evolving knowledge, industry needs, and political requirements.

There is much more of interest and substance about the varied trajectories of development in each of these resource sectors in the pages that follow, but let this précis suffice. Castonguay is an informed and spirited guide to these intricacies and to the role of technoscience in the elaboration and modernization of the Quebec state, its administrative capacities, and its territory. Other scholars, especially environmental historians, working on similar issues in other parts of the country might well bear this study in mind, not as a template but as inspiration. There is virtue in the combination of perspectives that Castonguay brings to bear in this work. Historical geography, environmental history, and science and technology studies have too often sought comfort in their own familiar silos, when they might, as here, have benefited from more adventurous interaction. Likewise, the scale and compass of this work is valuable not only for the insights it garners by considering several resource/economic sectors at the provincial scale but also for reminding us that ambition is a virtue. Transcendent, and useful, understanding of the large questions looming before us, about the conjoined fates of humanity and Earth that sustains us, will not come from timid inquiries.

In sum, Castonguay's analysis works on several fronts. Not least, it reveals the development of a robust and effective state apparatus in Quebec before the Second World War. The trajectory of these developments was of course unique, in detail, to Quebec – but it had its broad-form counterparts across the country. From New Brunswick to British Columbia, provincial governments ushered in new patterns and forms of resource management in the seventy years or so after Confederation. In this respect, Quebec was a province much like the others, one "entirely committed," in Castonguay's words, "to the national technoscientific enterprise." To adapt a slogan from the Quiet Revolution, the period of rapid change and modernization of Quebec society after 1960, the technoscientific agents discussed in these pages may not, in 1939, have been "masters of their own house" – but they had done much to familiarize themselves and others with the place, and had successfully erected a bureaucratic frame upon which subsequent transformations might be hung.[37] In telling this story, Castonguay gives us good reason – to borrow a phrase from Jones-Imhotep and Adcock – to think again about how "science and technology have formed the sites for Canadians to imagine, renounce, and reshape themselves as modern."[38]

Acknowledgments

THIS BOOK IS the culmination of research begun when I arrived at the Université du Québec à Trois-Rivières in winter 2002. What was originally conceived as a project focusing on the history and socio-politics of science changed and expanded following my graduate work on the autonomy of governmental scientific institutions. For a long time, I was interested in the history of the environment. I wanted to understand how nature was shaped by governmental scientific activity and, conversely, how the state's knowledge and power provided the basis for the potential for natural resources to be invented and inventoried, discovered and extracted as national assets. In this regard, my exchanges with my colleagues from the Network in Canadian History and Environment (NiCHE) were very stimulating. I would like to thank, in particular, my colleagues Matthew Evenden and Laura Cameron, as well as Liza Piper, for their encouragement. During scholarly events organized under the auspices of NiCHE, I had an opportunity to collaborate with geographers, from both Quebec and elsewhere, and to draw my most recent inspirations. As I looked more closely at historical and political geography, I sought to comprehend how the spatial dimension of human activity, both intellectual and practical, became an integral part of the environment I was studying.

As I made these detours, I benefited from the contributions of a generous group of people. I have to mention first the research professionals at the Centre interuniversitaire d'études québécoises (CIÉQ), to whom I owe a great deal, including the final quality of the composition of this book: Lauréanne Daneau, Philippe Desaulniers, Tomy Grenier, Jean-François Hardy, and Émilie Lapierre Pintal.

I would also like to express my immense gratitude to documentalist Lucie Comeau at the Centre interuniversitaire de recherche sur la science et la technologie, at the Université du Québec à Montréal, for having replied promptly to my numerous requests for material, much of which was buried in the meanders of the governmental apparatus. I also received assistance from archivists and librarians to whom I am indebted: Gordon Burr at McGill University, Monique Voyer at the Université de Montréal, Pierre Lavigne at the École polytechnique de Montréal, Richard Huet and Rénald Lessard at the Quebec City centre of the Bibliothèque et Archives nationales du Québec, and Christian Blais and Martin Pelletier at the Bibliothèque de l'Assemblée nationale du Québec. A great number of students and research assistants contributed to this project. I would like to thank, specifically, Guillaume Blanc, Gaston Côté, Olivier Craig-Dupont, and Darin Kinsey, as well as Jacinthe DeMontigny, Nicholas Toupin, Daniel Landry, Cathy Cloutier, Myriam Brouillette-Paradis, Louis Lacroix, Annie Carle, Charles Audet, Caroline Champagne, and Patrick Blanchet. Finally, my particular thanks go to Roger Bertrand, who, as I was beginning my research, gave me access to the results of his examination of the annual reports of a number of Quebec government departments.

The co-directors of the CIEQ, Yvan Rousseau and Marc St-Hilaire, agreed to publish the original version of this book in the "Géographie historique" collection; I am very grateful to them for this welcome and for their comments on the preliminary versions of the manuscript. My thanks also go to Guillaume Blanc and Michèle Dagenais for reading and commenting on parts of the manuscript. I am particularly happy to have undergone the collegial pressure of Michèle Dagenais and H.V. Nelles in favour of publishing this second "big book."

Over the years, I have received grants that enabled me to complete this project. I would like to thank the Canada Research Chairs Secretariat, the Fonds pour la Formation de chercheurs et l'aide à la recherche, the Social Sciences and Humanities Research Council of Canada, and the Université du Québec à Trois-Rivières for their financial support.

Some portions of this book were drawn from previous work, and I thank the directors of these publications for granting me reproduction rights: the *Revue d'histoire de l'Amérique française,* the Presses de l'Université de Montréal, and the Presses de l'Université Laval.

The Awards to Scholarly Publications Program made the publication of the English version of this book possible; the Canada Council for the Arts provided funding for the translation. I would like to thank the external reviewers who guided me in the revision of certain sections of the book. Maude-Flamand Hubert and Julien Prud'homme read the revised version, and they have my warmest gratitude for their availability and their comments. At UBC Press, James MacNevin and Ann Macklem made possible the publication of the English version, translated by Käthe Roth. I greatly appreciated their professionalism and their patience. And I sincerely thank Graeme Wynn for welcoming my book in the Nature | History | Society series.

The Government
of Natural Resources

Introduction

DURING THE NINETEENTH century, Western governments were deeply involved in technical and scientific activities aimed at intensifying exploitation of natural resources and stimulating industrial development.[1] Exploration, surveying, and mapmaking, which had been scientific activities at the forefront of territorial conquests in preceding centuries, acquired new strategic value in the wake of the Industrial Revolution. Following the French example, Great Britain and the United States financed expeditions to identify ore deposits likely to power the engines of manufacturing and the railway industry.[2] In addition to identifying coal and iron deposits, the geological and topographic surveys made it possible to organize the territory and communication infrastructure for emerging industrial economies. Like these Western powers, the Province of Canada was committed to defraying the salaries of geologists to discover deposits of coal to be used to maintain the colony's industrialization.[3] It also laid the foundations for a national statistical system and took astronomical observatories under its aegis to provide navigational aid.[4] The colonial state took many other similar initiatives to bring into its service individuals with scientific skills and techniques. Some, such as surveyors and wood cullers, joined the nascent public service. Others continued to work in civil society, including fish culturists, who operated their own hatcheries, and agronomists working for agricultural societies to spread agronomic information and develop techniques for eliminating insect pests or increasing crop yields.[5]

However, it was not until the late nineteenth century that the Canadian public administration undertook to systematically employ scientists to define the terms for exploitation of natural resources and occupation of the territory.[6] In the first decades following Confederation, a number of departments established scientific services to support the Canadian federal state's missions, such as the Meteorological Service in 1871 and the Experimental Farm in 1886. Initially, these services were run by naturalists with no formal training, but the creation of applied-science faculties and the emergence of research in Canadian universities, as well as the arrival in Canada of graduates from European and American universities, resulted in an influx of academically trained scientists to the civil service. Indeed, an observer of the time noted the pleasant scholarly atmosphere in the national capital early in the twentieth century.[7] Evidence of the transformation of Canadian universities and of state scientific services was the 1908 amendment to the federal statute governing the civil service to the effect that positions were to be filled according to the criteria of competency and merit, and no longer by virtue of patronage.[8]

Thanks to its financial resources, the federal administration was able to offer academically trained scientists permanent positions as civil servants in its different departments. Its constitutional responsibilities conferred taxation powers that enabled it to have a substantial source of revenue. At the same time, these responsibilities legitimized interventions in numerous areas, including some that were devolved constitutionally to the provinces. Initially less well endowed in terms of financial resources and human capital, provincial governments welcomed the federal government's incursion into their fields of jurisdiction, although they sometimes had to resort to the courts to protect their prerogatives. Their main lever was the administration of Crown lands, which, although a major source of revenue, raised concerns with regard to stewardship, especially because long-term availability of natural resources would be compromised if exploitation was left to private interests. Worse yet, information surrounding the availability of these resources was often deficient, which might negatively affect the possibilities of drawing revenues from them. This problem seemed even more serious because broad swaths of territory were being granted to entrepreneurs who were more interested in land speculation than in extraction of natural resources, thereby hindering the construction of a prosperous national industrial base.

Like its provincial peers, Quebec possessed little means and few people with scientific and technical training when it was created in 1867. The Canadian federal state had inherited the main elements of the administration of the Province of Canada, and the departments of the Quebec government had to limit their technoscientific activities to the collection and dissemination of information.[9] They turned to university professors and the federal state scientists to obtain the information they needed to formulate and implement their policies, sometimes funding local associations for the dissemination and application of knowledge. At the turn of the twentieth century, however, a number of departments set up scientific and technical services and acquired the capacity to hire individuals with academic credentials, at a time when the provincial public administration was undergoing pronounced growth.[10] Not only was the Quebec state now in a position to offer stable positions to university students enrolled in scientific and technical programs, but it was also undertaking the production of knowledge as its personnel travelled through the country to document and describe – through maps, surveys, plans, and reports – the province's riches and oversee their extraction.

The creation of scientific and technical services did more than sanction the instrumental function of science. Although the Quebec state was now endowed with *dispositifs* to frame the relations among the population, the territory, and the resources that it sought to "improve," the territorialities that its interventions produced necessitated a constant renewal of its knowledge base.[11] In addition to defining the terms for exploitation of natural resources and occupation of the territory, science had to produce its conditions of possibility within the state apparatus.

In this book, I explore the role of technoscientific activities in modernizing the intervention mechanisms of the Quebec state. For sectors related to the exploitation of natural resources, I reconstruct the activities of the scientific services and take a closer look at the integration of technosciences into governmental operations and at actions taken by scientists and technicians involving the territory and its resources. This portrait of the emergence and development of technoscientific activity in the Quebec public administration is intended to provide an understanding of some of the ways in which the state grew and functioned and to uncover how technoscience influenced appropriation and occupation of the territory, notably through practices that made the government of natural resources possible. Through this study of developments in mining and geology, timber cutting and forestry, hunting and fishing activities and wildlife management, and agricultural production and agronomy, I examine how exploitation of natural resources was built into a subject of knowledge and a tool of government, and I unravel the interweaving of two co-extensive phenomena: the production of state territorialities by technoscience, as well as the process of state formation and expansion of the state's administrative capacities.

The Territorialities of the Scientific State: Representations of Nature and the "Government of Men and Things"

Political geographers have moved away from their traditional focus on state borders and territorial sovereignty to become interested in the spatial practices of state institutions.[12] The construction and reconstruction of territories through which the state seeks to govern by "knowing and administering the lives and activities of the persons and things" remains, however, a largely ignored phenomenon.[13] For instance, historical geographers of social cartography have analyzed enumeration and statistical techniques without investigating the spatial foundations of these instruments of power.[14] Conversely, once their interest moved beyond the question of borders as determinants of international relations, political geographers studied the territorial bases of state power.[15] Among other things, for the state to be able to fulfill its fundamental functions – defence, order, and taxation – the representation of the space over which it has power becomes integral to its legitimacy in terms of affirming its sovereignty, ensuring the loyalty of its subjects, and protecting the ownership of its resources.[16] Too often, political geographers have limited their understanding of the relations between state and the territory to instrumentalization of the latter by the former, notably for regulation of the circulation of goods, individuals, and even ideas.[17] The power of the state, from this perspective, resides in its capacity to define a space and circumscribe the mobility of people and things to that space. However, the conditions for shaping the territory in order to instrumentalize it remain obscure.

Under what conditions, then, is the construction of a territory capable of building state power? The geography of territoriality proposes two complementary approaches to this question.[18] In the view of Robert Sack, territoriality is understood on the basis of the strategies that social actors articulate to influence and control resources and populations by defining the space upon which their jurisdiction is exerted.[19] Claude Raffestin sees territoriality, rather than simply the result of a deliberate strategy, as Sack theorizes, as a system that produces relations between social groups and material environments, which are modulated by mediators (instruments or representations, for example).[20] Raffestin asks us to appreciate the indeterminate nature of territorial production and to examine the significance of territories and the actions that produce them. Although Sack's approach targets mainly the outcomes of territorial production with regard to both objectives and results of territorial practices, it has the merit of emphasizing the asymmetry of power relations in a relational system, notably for territories whose functionalities and materialities endure to the point of rigidifying the borders and characteristics of contested spaces.[21]

To understand the relations between state and territory, I propose, following both Raffestin and Sack, to investigate how these two entities are mutually constructed.[22] Here, it seems instructive to explore how historians of science address the territorial enterprise of a state that, through its scientific activity, seeks to produce and govern spaces to affirm its sovereignty. Notably, analysis of national – and nationalist – maps has made it possible to consider the terms of construction of a statist territory through the sciences that define its contours.[23] Cartographic projects have therefore been the subject of numerous analyses that have revealed the normative dimension of this descriptive scientific activity.[24] Rather than an objective a priori that the cartographer would simply reproduce, the correspondence to reality of a map results from the material shaping of the territory made possible by actions undertaken on the basis of that very cartographic representation. It is as if one were representing less what is there than what one wishes to represent or show: "A map anticipated a spatial reality, not vice versa. In other words, a map was a model for, rather than a model of, what it purported to represent."[25] Finally, a map performs an eminently nationalist function, in that it aims to lend unity and coherence to a territory the cartographic representation of which is intended to disseminate to populations otherwise dispersed throughout a series of discrete spaces.[26] A national map, however, does not just provide an image of a state in search of legitimacy, but functions above all as a means by which such an entity may effectively be imagined, propagated, and circulated.[27]

Scientific production of a territory goes beyond cartography to include surveying activities that provide a description of the territory, assigning to it productive functions related to the human and natural resources enumerated.[28] The territory that is represented is intended to be rationally organized to both serve the needs of the state and facilitate occupation of that territory. Two spatial dimensions of scientific activity must be underlined. First, surveying activities proceed from a territorial undertaking because they require demarcation of that territory. Once the qualities of a territory are enumerated – for instance, soil fertility, geological formations, ore deposits, or plant or animal species – the state can define spaces, either to put them to a particular use or to protect resources for future exploitation or for conservation, often for the privilege of an exclusive group of users.[29]

In addition, the production of inventories presumes the circulation of state agents – explorers, surveyors, census takers, naturalists – through the territory to define borders, evaluate assets, measure and assign lots, and trace out and open roads. Like land agents responsible for allocating timber limits and collecting revenues on public lands, explorers mark out the space by their presence, as brief as it might be. Indeed, as representatives of the state, these civil servants are invested with a certain legitimacy in the definition of uses and users of the territory and its resources. Aside from explorers, the definition of larger or smaller parcels of land, in the form of timber limits, reserves, or parks, requires the deployment of state agents who, endowed with a degree of legal authority, are responsible for protecting the functions of the territory.[30]

Although historians and geographers have emphasized the role of surveying and mapmaking in the rise of the modern state, the state's authority is also based on the material organization of the territory.[31] In this respect, science acts not solely through the force of its ideas but also because putting those ideas into practice gives form to an environment that participates in the "government of men and things" by making these entities visible.[32] At the core of the materialization and shaping of the environment are legibility practices used by a state to make up for the disorderliness of nature and society, which contradicts the representations of its technoscientific and political enterprise.[33] Whether by regulation of extractive practices or improvement of the environment, the state inscribes on the landscape the standards and norms formulated by its personnel to oversee access to the territory and use of its resources. Similar work contributes to the simplification and normalization of the territory, and its legibility is increased by the distancing or smoothing over of what would otherwise be considered irregularities, ordinarily attributed to uncontrolled natural phenomena or resistant populations.[34] In this respect, technoscientific activity plays a decisive role: the elements of the landscape that it makes visible become targets of state intervention, opening the potential to regulate occupation of the territory, thereby facilitating extraction of natural resources and increasing state revenues. Similarly, through the naturalization of categories proposed by the state apparatus, technoscientific activity enables the legitimation of the social order on which this very apparatus depends.[35]

State Formation: Administrative Capacities and the Government of Conducts

How do technoscientific activities that are spatial practices, such as cartography and surveying, the circulation of civil servants, and improvement of the environment, reinforce state power? By representing and intervening in the territory and its natural resources, the state manifests its capacity to govern not only an environment but also its populations, human and non-human, to regulate their conducts, and to circumscribe, through human or non-human entities, their freedom to act. In the exercise of this action at a distance resides the government of natural resources. With the accumulation of knowledge, the acquisition of qualified personnel, the establishment of infrastructure, and the deployment of *dispositifs* in the field, the state expands its administrative capacities. It is built on a territory that it governs at the same time as it creates it, by delineating it and defining its forms and linkages.

Both the territory and natural resources constitute objects of knowledge and targets of political intervention that the state invests with representations and infrastructure, thus intruding in the landscape and society.[36] Does this mean that the state controls the population and the territory? The issue seems to be less about state power and more about the strategies that shape "practical objects" for governmental interventions. What matters is not outright control but regulation of access to and uses of the territory and its resources.[37] Furthermore, as anthropologist James Scott shows, the simplifying maneuvers of states are often circumvented by local populations that mobilize "*metis* knowledge" to subvert the natural and social order that the political and technoscientific authorities seek to establish.[38] Therefore, one needs to understand how populations participate in this "order of things" in order to identify the territorial strategies upon which the state depends to augment its political power.

A similar perspective is pervasive in studies on governmentality that examine the state less as an autonomous institution than as the exercise of power grasped analytically and expressed relationally. Governing does not simply mean controlling. On the contrary, a relational analysis of power recognizes that individuals have the freedom to act within limits predefined by knowledge that participates in the "normation" of conducts.[39] In this perspective, a central power is not said to spread its influence through society by expanding state control techniques. Furthermore, to explain the governmentalization of the state – the means by which the administrative apparatus undertakes to understand and manage the lives and activities of people within a territory – these studies posit that the state takes over fields of power and social groups that are already governing – that are involved in "shap[ing] and administer[ing] ... the lives of individuals in pursuit of various goals."[40] Understanding power and its effects, which are negative or productive depending on whether it simply oppresses segments of the population or generates capacities for action among individuals, thus requires taking into account a multiplicity of self-governing actors.

Although governmentality studies seek to understand how the state aims to transform the density and vitality of certain territories and to make their populations productive, the land and its riches remain concretely and materially absent from these reflections.[41] Spatial and political theorist Stuart Elden reminds us, however, that statistics as a technique of government is based on spatial distribution. Because it ceases to be a static terrain and becomes "a vibrant entity, with its 'specific qualities' which too can be measured," territory is central to governmentality, in Elden's view.[42] This perspective provides an understanding of how the qualities and the geometric and relational properties of the territory are the subject of calculational strategies (mapping, ranking, measuring, defining, normalizing, regularizing, networking) similar to those employed to regulate populations. Governing, knowing, and regulating the population and the territory flow from a single political rationality, the territory being understood in its relations with the population.

Geographers have recently emphasized the need to incorporate space into our understanding of the government of conducts[43] and to comprehend territory as political technology, drawing from the writings of philosopher Michel Foucault an illustration of the articulation between territory and governmentality.[44] Foucault shows that security mechanisms formulated in the nineteenth century, which succeeded those implemented to discipline bodies or police and control territory, were based on organizing the environment to facilitate "liberal" circulation in the perspective of the "government of men and things." What a *dispositif* problematizes is the environment *(milieu)* – the space of conducts – that the state seeks to organize in order to govern populations starting from the identification of regularities. Governmentality is thus distinct from sovereignty and discipline. Each in its own way, these political rationalities have made the exercise of power possible, the former by targeting "a set of legal subjects capable of voluntary actions," the latter by taking on "a multiplicity of organisms, of bodies capable of performances."[45] Rather than targeting the territory and the subjects living in it, as sovereignty does, or training and correcting the body of the individual, as discipline does, governmentality is exerted over a population, "a multiplicity of individuals who are and fundamentally and essentially only exist biologically bound to the materiality within which they live."[46] Territory and population, indivisible, are concurrently the subject of problematizations by technoscientific activities that expand the administrative capacities of the state both through the *dispositifs* that they deploy and through the government of conducts that they engender, or through the mechanisms of knowledge that they elaborate to make these entities visible and to act on them.[47] Following a Latourian epistemology, post-humanist geographers even ask us to consider "men and things" necessarily and together: nature and society form hybrid networks, integrating processes that are both human and natural, both mechanical and organic.[48]

The Territory of the Quebec
Public Administration and Its Riches

In this book, I examine the technoscientific activities of the Quebec state that accompanied the development of economic sectors linked to the exploitation of natural resources. My goal is to specify the role of these activities in the functioning of the administrative apparatus between 1867 and 1939. Far from being confined to the margins of the state apparatus, where their role would be limited to data collection, technoscientific activities were a driving force in modernizing the mechanisms of state intervention. Furthermore, the territorialities that they shaped during the formulation and implementation of state interventions contributed, in their turn, to the formation of the state. By actively defining new terms for occupation of the territory and exploitation of natural resources, technoscientific activities thus supported the spatial and political expansion of the state.

The territory of the province of Quebec has been the subject of numerous studies by geographers and jurists, whether to establish the conditions of its occupation or to account for the delimitation of its borders.[49] In the half-century that followed the creation of the province of Quebec, the territory remained fluid, due to the proliferation of settlement frontiers and the indeterminate status of its legal boundaries. The boundaries were extended first in 1898, when the federal government ceded to the province – at the time defined by the borders of Lower Canada (Canada East) – the northwest district of Abitibi, and removed a part in the northeast, on the coast of Labrador. Then, in 1912, the northern district of Ungava was added. In both cases, changes to the territory of Quebec resulted from actions taken by the governments in London and Ottawa. Quebec governments played a more active role in the shaping of the territory by financing construction of the railway and colonization societies, even though initially it was the Catholic Church authorities that carried out this latter form of territorial expansion.[50] As the old settlement parishes in the Laurentian plain became overcrowded, people struck out to conquer the Appalachian and Laurentian plateaus in the first quarter of the nineteenth century. In the second half of the century, settlement began to expand to Témiscouata, Témiscamingue, and Lac-Saint-Jean, regions with a climate and soil more favourable to farming than the Canadian Shield. After the addition of the Abitibi and Ungava districts to the territory of the province, urban settlements sprang up in these regions, thanks to the economic activities spurred by the extraction of mineral resources.[51]

Few authors have seen the territory of the province of Quebec, with its shifting contours, as a subject propitious to analysis of the role of knowledge, legal or scientific, in the functioning of the state and the evolution of its administrative apparatus. Jurists have discussed the legal and historical difficulties behind a precise definition of the territory of Quebec, but such analysis implies that it would be possible to remove all ambiguity behind comprehension of what this territory is and should be as a space for exercising exclusive jurisdictions.[52] Geographers have explored how different phases of settlement gave rise to differentiated occupation of the territory in terms of the full or partial settlement of regions, the activities that took place in them, and the creation of distinct landscape formations.[53] More rarely have scholars addressed interventions by the Quebec government linked to exploitation of natural resources and occupation of the territory. One exception is the historical atlas aptly titled *Le territoire*.[54] The contributors to this book are interested specifically in territorial formations related to the mining and forestry industries, fish and game activities, and agricultural modernization, and they emphasize the impacts of these activities on the landscape, their articulation with a constantly evolving urban space, and their integration with the Laurentian ecumene and the continental economy.

There is, however, an abundant literature in economic and social history on natural resources industries in Quebec, and governmental policies are often mentioned.[55] People employed in the mining, forestry, and fishery industries have been of interest to labour historians studying the struggles and living conditions of workers, notably in the towns and villages created for them to live in, which were to shape new communities.[56] Historians have also examined the fish and game clubs, analyzing the social relations underlying exploitation of the wildlife resources and of society by elite sportsmen.[57] Finally, given the significant amount of capital necessitated by extraction of natural resources, business historians have paid attention to companies and the relationships that they maintain with governments and local communities.[58] Although they are more interested in the socio-political landscape than the natural environment, social and economic historians of natural resources industries have highlighted social inequalities and labour struggles that have had consequences for development of the territory. Moreover, they have shown a certain sensitivity to the material dimension of the territory and to resources and their inscription in the regional imagination.

In general, historians focus on a single sector, without necessarily trying to investigate how state interventions and territorial transformations underlying the exploitation of different natural resources follow a similar dynamic. Furthermore, when historians investigate state interventions related to resource exploitation, their studies are usually limited to an assessment of the impacts on the economic development of the province. Sometimes, they address the lives and actions of chief scientists or directors of science services, but without paying due attention to how technoscientific activities, governmental policies, and regulatory measures play out in the field, especially with regard to how they shape the landscape or the pace of resource exploitation. Transversal studies of these sectors are rare, especially when it comes to understanding the formation of territorialities and state interventions.[59]

It must be said that historical research on the Quebec state, especially for the period covered here, is generally limited to denouncing its lack of financial resources and the absence of capable civil servants, with a concomitant incapacity to engage in economic activity other than by selling off its natural resources. After a lean period following Confederation – punctuated with short intervals of prosperity but characterized mainly by the Great Depression of 1873 and repeated economic crises – the Liberal regime that ruled uninterrupted from 1897 to 1936 made influxes of foreign capital the cornerstone of the province's industrial development.[60] This strategy of support for exploitation of natural resources was based on promoting and publicizing the province's "natural riches," as well as granting timber limits, mining claims, and hydroelectric sites – which nevertheless remained in the public domain – at rock-bottom prices.[61] The Liberal governments left the path entirely open to private initiative, as had preceding governments, most of them Conservative, that had led the province since 1867. Whereas the Conservative governments counted on railway construction and modernization of agriculture to stimulate settlement, integrate remote regions, and keep the French Canadian population within the province, the Liberal governments of Félix-Gabriel Marchand, Simon-Napoléon Parent, Lomer Gouin, and Alexandre Taschereau were more interested in stopping emigration to the United States by welcoming foreign investments. For these Liberal premiers, foreign investments guaranteed the creation of factories and industrial jobs and helped to increase state revenues, which depended in large part on royalties from extraction of natural resources and leases on vast stretches of land.[62] Contemporaries criticized this strategy, loudly accusing the government of selling the province at a discount to foreign capital, for which they would have preferred to substitute indigenous capital – although its availability remained uncertain.[63] They also denounced the fact that many ministers and premiers in Liberal governments were, conveniently, sitting on the boards of directors of major corporations that benefited from the largesse of the Quebec state.

Aside from providing studies of a few political figures, the historiography portrays the Quebec state of this period as non-interventionist, poorly coordinated, and under-funded, with Ontario often presented as the contemporary example of a modern, efficient public administration.[64] The Quebec public administration thus appears to have operated with an absence of logic and planning; its fundamental approach to economic and social affairs was apparently based on laissez-faire and omnipotent private enterprise capable of manipulating the state's levers.

A few remarks are in order. First, even in a laissez-faire political system, the state remains responsible for "the structures and regulations that made such free action possible."[65] Therefore, in the nineteenth and early twentieth centuries, governments that claimed to subscribe to economic liberalism, in Canada and elsewhere, multiplied their interventions in economic affairs. They both instituted a regulatory framework to guarantee protection of and respect for private property and funded, partially or entirely, the construction of roads, canals, and railways to facilitate circulation of goods and people.[66] The study of governmentality is interesting precisely because it seeks to understand the means implemented by the state to resolve the dilemma of "governing less to govern better."

Second, it must be noted that the history of the Quebec public administration is, in many respects, similar to that of the other Canadian provinces. In Ontario and British Columbia, to mention only two, the functioning of the public administration was initially characterized by the absence of financial control and the incapacity to enforce laws, as well as by insufficient human resources, not to mention the fact that those who obtained positions in the civil service did so through the dubious mechanisms of patronage and political partisanship.[67] These two provinces began to modernize their bureaucracies only in the early twentieth century by, among other things, instituting civil service commissions and appointing professionals on the basis of their competency.[68] In this, they were following the example of the federal administration, which founded its Public Service Commission in 1908. It should also be noted that research on the economic history and the history of public administration of Quebec has shown that the interventionism of the Quebec government compared favourably with that of Ontario well before the onset of the Quiet Revolution.[69] Notably, Quebec invested in occupation of the territory and exploitation of natural resources in the decades that followed Confederation. And it is precisely in these sectors that technoscientific activity, whose terms and scope remain to be elucidated, took shape.

Finally, even if for the period under study laissez-faire and private enterprise dictated governmental interventions, the state had to have intellectual resources to respond to this demand. Where did they come from? How did they act? What were their "effects of power"?[70] Depending on the sector under study, the Quebec government may appear to have been less well coordinated and more underfunded than the Ontario government. In hydroelectricity production and distribution, for example, the early nationalization of this industry in Ontario encouraged more interventionism and the development of expertise for expansion in the sector.[71] Similarly, discoveries of ore deposits around Sudbury and in northeastern Ontario in the late nineteenth century led the provincial government to invest immediately in the mining sector, first by creating the Bureau of Mines in 1891, then the School of Mining in Kingston in 1893.[72] Although it was founded in 1883, the Quebec Bureau of Mines began to expand only forty years later. That the early creation of institutions was not a guarantee of sustained interventions is also evidenced in the forestry sector, but to the disadvantage of Ontario this time. That province's public administration was granted a bureau of forestry in 1898 and a first professional forest manager in 1905, but it increased its personnel to supervise the exploitation and development of the forestry sector only after the Second World War.[73] In contrast, as soon as the Quebec forestry service was founded, in 1909, it began to undertake silvicultural work.[74] Finally, exploitation of wildlife resources in Quebec was also extensively supervised by agents of the provincial administration, whereas Ontario seems to have relied more on the federal administration.[75] Therefore, to determine the terms and scope of the state's technoscientific activities, we must consider the conditions related to the exploitation of each resource and the institutional ecology of education and research in each sector.[76]

How This Book Is Organized

In the five chapters of this book, I examine the conditions under which the state of Quebec and its territory were constructed through the deployment and activities of its technoscientific agents. My approach is based on an analysis of the institutional ecology of the scientific institutions in the state apparatus and on a political and historical geography of governmental interventions by the Quebec state around the exploitation of natural resources. I illustrate how technoscientific activities participated in the formation of territorialities and the growth in administrative capacities by and for the government of natural resources. I focus on the activities of civil servants as they formulated and implemented programs and policies, yet I do not lose sight of realities in the field, where those outside the public administration were active in exploiting natural resources, limiting their accessibility to facilitate or impede extraction, or protecting and conserving them for future use.[77]

The period covered ranges essentially from 1867 to 1939, even though I sometimes turn back to developments in the Quebec public administration just after the Province of Canada was established, when institutions were created for the production, application, and dissemination of knowledge likely to stimulate the expansion and industrialization of the British colony. The analysis concludes in 1939, on the eve of the Second World War. Radical changes to organization of the sciences within the federal administration during the war had repercussions for scientific institutions across the country in the following decades, including within provincial administrations.[78] Other factors militate in favour of this upper time limit, which was a time of endogenous scientific development within the Quebec state apparatus. First, a number of provincial scientific services undertook the construction of laboratories in Quebec City. Second, a vast survey was undertaken under the aegis of economist Esdras Minville, a professor at the École des hautes études commerciales in Montreal; this operation mobilized academic and state scientists to take stock of the exploitation of agricultural, forestry, mining, fishery, and wildlife resources.[79] These developments were another manifestation of the "vire d'un temps nouveau" in which, under the Union nationale government elected in 1936, the French Canadian scientific community finally found the political authorities turning an attentive ear to its demands for consolidation of its institutions.[80]

In Chapter 1, I provide a transversal overview of the development of governmental services linked to the exploitation of agricultural, forestry, mining, and wildlife resources; the recruitment and training of specialized personnel; and their activities within the public administration.[81] This overview offers an opportunity to describe the ecology of educational and scientific institutions on exploitation of natural resources in Quebec, and to shed light on the organization and operations of a technoscientific bureaucracy. Here, I examine the formation of the state in its literal sense: the administrative apparatus grew in size, budget, and personnel in order to engage in various technoscientific activities, with the creation of governmental services and the implementation of measures affecting exploitation of natural resources and occupation of the territory.

In the next four chapters, I explore the role of technoscience in the expansion of the administrative capacities of the Quebec state, focusing on how scientists constructed territorialities through which national riches were to be located and extracted. My aim is not to provide highlights of state science in Quebec but to use case studies to demonstrate the different conditions for production of territorialities through which technoscientific activity contributed to the formation of the state. In addition to drawing on textual and iconographic analysis from the printed cartographic and archival sources in the collections of state and university institutions and industrial associations, I use data from the public accounts and departmental annual reports to decipher the spatiality of the technoscientific practices of civil servants for the government of natural resources.

In Chapter 2, I address the efforts of the Quebec state to stimulate mining development. The Geological Survey of Canada had been reporting and inventorying mineral resources in Quebec since its foundation in 1842; some forty years later, the Bureau of Mines of the Quebec Department of Crown Lands undertook the exploration of the province's territory. The provincial government itself had to set the priorities for this work and direct the explorations where it seemed possible to establish, if not a mining operation, at least settlements on the newly acquired lands. The idea was to take over a territory not simply by describing or defining it, nor even by mapping it, but by probing the subsurface with a view to extracting its riches. Whereas maps from the Geological Survey of Canada indicated only rock formations, the Quebec Bureau of Mines turned out to be more concerned with the cartography of mining centres in the province, both real and potential. The maps produced by geologists from the Bureau of Mines no longer portrayed natural geological entities in continuity with the rock formations of other Canadian provinces – which, incidentally, integrated Quebec into a pan-Canadian whole. Instead, they showed the territory of the province as a coherent set of mineralogical resources to be exploited and toward which the industry was to be directed. Industrial and economic concerns influenced the bureau's approach during its early years, when it lacked specialized personnel. Later, the addition of geologists and the creation of geological and cartographic divisions within the bureau led to the performance of fundamental work that, by inventing a mining space, oriented both exploitation of the province's underground riches and detailed exploration of the mineral belts.

The strategies of the Department of Crown Lands with regard to the leasing of forestland and, in parallel, the management of conflicts between forestry businesses and the colonization movement are the subject of Chapter 3. One of the first tools used by the Quebec government to keep settlers away from the forest was the creation of reserves and parks. These territories, delimited to protect trees as a resource over the long term, were added to the forest regime in force since the mid-nineteenth century, based on the awarding of timber limits to companies for the immediate cutting of wood. In addition to delimiting timber limits, the surveyors of the Department of Crown Lands classified the territory to separate land for settlement from forestland deemed inappropriate for farming. At a time when a pulp and paper industry was emerging that was both voracious and less selective with regard to species of trees consumed, the introduction of scientific forestry in the early twentieth century had the effect of putting the forest regime on a different footing. As a consequence, relations between settlers and the forestry industry were to be redefined. The Forest Service of the Department of Lands and Forests, which was also responsible for creating and running the forestry school attached to Université Laval, made reforestation a tool for training its personnel and providing settlers with wood through township reserves. Thanks to their location in settlement areas, as well as to the tree species used for reforestation, these reserves grew a pulpwood forest for the paper industry. This shaping of the landscape was a supplementary step in the separation of the land for the agricultural and forestry sectors and functioned to distance settlers from the territories allocated to holders of timber limits.

In many respects, Chapter 4 illustrates how the exploitation of wildlife resources was based on an approach similar to that used for the forest: creation of reserves and improvements of Crown lands. Yet, surveillance was central to two additional governmental strategies based on the regulation of human and non-human populations: access to the territory for fish and game clubs and fish stocking through pisciculture activities. The commissioner of Crown lands proceeded to lease out vast domains to private fish and game clubs in exchange for an annual rent, to which sportsmen also added royalties for permits and catch. Once leased, these lands became spaces where the clubs enjoyed exclusive catching rights, often to the detriment of other users. Conflicts arose between Indigenous people and Euro-Canadian settlers, on the one hand, and elite sportsmen who wandered through the backcountry to harvest game and fish that the "locals" used for food or trade, on the other hand. Although successive governments justified such arrangements by arguing that they would otherwise have to spend money for the stewardship and protection of wildlife, they nevertheless provided financial assistance to private fish and game clubs for the surveillance and improvement of the leased territories. Fish and game wardens travelled through the country to regulate human populations involved in exploitation of wildlife. Similarly, the Fisheries and Game Service recruited biologists to oversee regulation of animal populations. Through its hatcheries and pisciculture activities, the service was able to shape the aquatic landscape so that sufficient numbers of fish were available to maintain the attractiveness of Quebec's watercourses as a "sportsmen's paradise" to welcome populations of fish anglers from abroad.

In Chapter 5, I examine governmental interventions in agriculture. During the period under study, the state shaped an agronomic space through regionalization of agricultural production and the formation of subjects – commercial farmers – as the development of dairy husbandry and the adoption of other kinds of specialized production were bringing the agricultural economy out of the doldrums. In comparison to the three sectors studied in previous chapters, research in agriculture started relatively late, even though the Department of Agriculture opened a laboratory for dairy control and soil analysis in 1888. Although technoscientific interventions in agriculture revolved mainly around education and extension activities, they nevertheless shaped the agricultural landscape over the long term, notably after the arrival of "official agronomists." Once the Department of Agriculture had hired graduates from the agricultural schools of the province's universities, dissemination of agricultural knowledge rested on the interventions of civil servants rather than members of civil society. Although the agronomic space reproduced the territorialities of political representation and the boundaries of electoral ridings, the activities of the Agronomic Service left an imprint on the agrarian landscape, notably through the formation of specialized producers and the geographic anchoring of agricultural specialties such as fruit growing and poultry production.

I
The Administrative Capacities of the Quebec State: Specialized Personnel and Technoscientific Interventions

IN THE LATE NINETEENTH century, the Quebec state began to hire scientists, up to then almost entirely absent from the provincial administration. It must be said that, in general, the number of employees in the civil service was small; the administration had eight departments in 1896, one more than in 1868, and each had an average of twenty-five employees, most of them clerks.[1] Notwithstanding the ambient economic liberalism, which discouraged governmental intervention and expansion of the administrative apparatus, the Quebec government had a narrow margin of financial maneuver in the decades following Confederation. Discussions surrounding the sharing of the debt of the Province of Canada created uncertainties that limited Quebec's capacity to borrow, and in 1873, when the federal government decided to assume responsibility for the debt, the international economy was seriously shaken by the onset of the Long Depression. The unfavourable economic conditions damaged the investment climate for years thereafter, dragging down the already reduced revenues that the provincial treasury could count on obtaining.[2] As a weak recovery took hold, provincial departments set up their own services to fulfill their responsibilities related to administration of natural resources and occupation of the territory, such as the Bureau of Mines and the Fisheries Service in 1883 and the Fisheries and Game Service in 1885. When, between 1896 and 1913, Canada experienced a period of strong growth, Quebec was able to extract more revenues from exploitation of natural resources on public lands and extend its control over various sectors. New departments appeared, notably following the transformation of the Department of Crown Lands, which oversaw forests, mines, fisheries, and hunting, and the Department of Agriculture and Colonization. The state took on new missions through complexification of its organization and the hiring of civil servants, whose numbers rose from 526 in 1899–1900 to 5,745 in 1930–31.[3] Created in 1905, the Department of Lands and Forests established a forestry service in 1909, and an agronomy service was created within the Department of Agriculture in 1913. The Department of Colonization, Mines, and Fisheries absorbed the Bureau of Mines and the Fisheries and Game Service, to which it added a hatchery service in 1916. During the decades that followed, these services diversified their activities not only because financial and human resources grew but also because scientific knowledge became more specialized, governmental programs multiplied, and the regulatory apparatus was strengthened. A central element of this diversification was the availability of a recruitment pool of qualified workers thanks to the transformation of institutions of higher learning.

In this chapter, I present an overview of these organizational developments and the institutional ecology of technoscientific activity in Quebec. For the departments related to agricultural, forestry, mining, and wildlife resources, I describe how the scientific services developed by analyzing the recruitment and activities of qualified workers. Based on data drawn from the public accounts and annual reports of the departments concerned, this description provides a glimpse of the extent of the administrative capacities of the Quebec state. First, I examine how hiring was conducted by looking at the training available and the needs of the administrative apparatus. Among other things, I describe the changes in institutions of higher learning that encouraged the training of qualified personnel in sectors related to exploitation of natural resources. Then, I look at the various interventions by scientific services at different times and their effects on the operation and organization of departments in the sectors concerned.

Growth in Public Administration: Scientific Personnel

The late nineteenth century initiated a period of prosperity, at the very time when economic development in Quebec was building on the Second Industrial Revolution and the increasing value of "new" national riches. Just like milk and freshwater salmon *(ouananiche)*, hydroelectric power, asbestos, and pulpwood were resources whose qualities or means of extraction and processing were being updated by technoscientific advances, and this enabled Quebec governments to change their view of the province's territory and assets. Also during this period, the state administration became more specialized. After the Liberal government of Félix-Gabriel Marchand came to power in 1897, the province's commissioner of agriculture shifted activities related to colonization to another department, leaving him free to devote himself to reorientation of the agricultural industry around dairy production. The Department of Crown Lands was divested of the different sectors of natural resources that it had originally covered. Relieved of responsibility for mining, and then for hunting and fishing, by 1905 it was concerned only with forestry operations and administration of Crown lands.[4]

Scientific personnel attached to the internal service – mainly permanent employees working full-time at the seat of government in Quebec City – grew modestly before the beginning of the twentieth century, but there were two major surges between 1897 and 1937 (see Table 1.1). The small number of scientists in the internal service at the beginning of the period under study was the result of limited interventionism; this, in certain respects, reflected the limited responsibilities of the provincial government, which counted on outside agents to gather and distribute information. Another cause of low staff numbers may have been the state of the public treasury, which was strained by debt servicing related to major loans incurred to finance the construction of railways, and by decreased revenues, both from exploitation of natural resources, which had slowed since 1874, and from federal subsidies, which no longer matched the province's demographic situation.[5]

A first wave of expansion in technoscientific personnel took place in the early 1910s. In addition, a change to the classification of public servants in Quebec in 1912 probably resulted from the growing number of employees involved in technoscientific activities, as a new category was established to recognize the "special professional, scientific, or technical knowledge" of certain employees.[6] The second wave of expansion began in the mid-1920s.

TABLE 1.1 Technoscientific personnel in the internal service

Year	Agriculture	Forests	Mines	Fish and game
1897	4	10	1	3
1902	4	10	1	3
1907	4	10	2	3
1912	10	19	2	3
1917	13	22	2	4
1922	12	22	3	4
1927	49	27	3	4
1932	73	40	8	4
1937	92	59	13	7

During the first series of hirings, teaching institutions founded with government funding – the École de l'industrie laitière du Québec (dairy industry school of Quebec) in Saint-Hyacinthe, in 1892, and the École de foresterie (the Quebec Forest School) at Université Laval, in 1910 – played different roles in their respective sectors.[7] Soon after the teaching of forestry, which began at the Berthierville tree nursery in 1907, was integrated into the École de foresterie, the Forest Service experienced regular increases to its personnel, already quite numerous at the end of the nineteenth century.[8] The main tasks of this service were surveying, exploration, and, to a lesser extent, conservation of the forest to prevent overexploitation and destruction by fire. In contrast, the creation of the Agronomic Service in 1913 involved only the appointment of a director, and there was no increase in personnel of the internal service despite the diversification of responsibilities of the Department of Agriculture. Whereas forestry graduates of Université Laval found openings in the scientific service, what seems to have taken precedence at the École de l'industrie laitière was training in supervision subsequent to strengthening of the regulatory apparatus and the creation of inspection services to apply the regulations.[9] After speakers and editors of the department's main information organ, *Journal d'agriculture illustré*, inspectors general for cheese and butter factories were the first specialized positions created in the internal service, in 1912. These employees also taught at the École de l'industrie laitière, where inspectors and factory employees were trained and certified.

Whereas the first wave of recruitment was manifested in the forestry and agriculture sectors alone, all sectors benefited from increased staffing in the internal service during the interwar period. This was particularly true in the late 1920s, after a decade of prosperity during which exploitation of natural resources intensified. The second wave of hiring in the Department of Agriculture testified to the proliferation of agricultural services and specialties, after decades of efforts devoted to developing animal and horticultural production, particularly in the dairy, beekeeping, and fruit-growing sectors. It was at this time, and for reasons that I explore later in this chapter, that the scientific personnel in the Department of Agriculture expanded with the creation of the Agronomic Service and the hiring of graduates from agriculture schools at Quebec universities. In the forestry sector, after the initial growth in the number of staff positions for surveying, inspection, and classification of the territory, tasks related to forest conservation stimulated recruitment by the state apparatus.[10] The Department of Lands and Forests was then expanding its responsibilities with regard to reforestation and timber inventory.[11] The many new hires in the late 1920s resulted from increased specialization of work related to administration of the forest estate, as the department created positions expressly to conduct silvicultural, meteorological, entomological, and botanical studies and to continue with forestry management.[12] Recognition of scientific specialties was also seen in the mining and wildlife sectors, but not until 1929. In that year, the Bureau of Mines created its Division of Geology and hired specialists in geology, geodesy, and cartography. Similarly, the Fisheries and Game Service recruited in 1930 a biologist who had graduated from McGill University, Bertram William Taylor, to oversee pisciculture activities. Up to then, amateur naturalists had occupied the positions of superintendent and inspector of fisheries and game since the service was created in 1885.[13]

Whereas technoscientific personnel grew modestly in the internal service before the mid-1920s, it was different in the external service. Despite slowdowns registered during the First World War and the economic crisis of the 1930s, growth in personnel outside the seat of the provincial public administration in Quebec City was initially constant, then exponential, starting in the 1920s (see Table 1.2). This is where we must look to have an idea of the extent of scientific activities, many of which were reflected in the creation of seasonal jobs usually dispersed throughout the territory, as a function of resource-extraction cycles. It was similar for employees whom the administration hired on a permanent basis but for whom it was slow to regularize positions.

The number of speakers, instructors, and inspectors grew considerably in the provincial Department of Agriculture after the federal government adopted the 1913 Agricultural Instruction Act.[14] The statute, which was in force for ten years, provided for the payment of grants to the provinces, which were constitutionally responsible for education, to promote agricultural training. Under this law, the federal government transferred $2,640,000 to

TABLE 1.2 Technoscientific personnel in the external service

Year	Agriculture	Natural resources
1897	14	71
1902	18	104
1907	18	157
1912	36	129
1917	146	136
1922	171	254
1927	148	371
1932	174	311
1937	197	352

the Quebec government, as well as $20,000 annually for veterinary training, between 1913 and 1922.[15] This money was used for purposes as diverse as the creation of a veterinary school, the appointment of a provincial entomologist, and the hiring of agronomists and instructors. A number of seasonal positions created ended up being integrated permanently into the internal service in the 1920s, after Robert Borden's federal Conservative government decided not to extend the law in 1923. In the meantime, the Quebec Department of Agriculture added many new positions for spreading knowledge within the Agronomic Service that it had created in 1913.

This growth in scientific personnel benefited from the creation of agriculture schools in Quebec universities not long before. After the foundation of Macdonald College in 1906, McGill University had an agriculture school, and Université Laval and its Montreal branch affiliated with, respectively, the École d'agriculture de Sainte-Anne-de-la-Pocatière, in 1912, and the Institut agricole d'Oka, in 1908.[16] The first bachelor's degrees in agricultural science were awarded in 1911 and 1912 at McGill University and Université Laval, both of which also had a veterinary school; there were in fact three veterinary schools in the late nineteenth century, although the one at McGill closed in 1913.[17] These educational institutions made it possible for the Department of Agriculture to hire agronomists on a regular basis; their numbers rose from 5 to 68 between 1913 and 1926, and then from 78 in 1931 to 388 in 1943 (see Chapter 5). When he announced the appointment of the first five official agronomists in 1913, the minister highlighted the fact that they had all graduated from the Institut agricole d'Oka.[18] Their role was to transmit information on agronomy to the farming community through lectures, exchanges of correspondence, and demonstrations.[19] Employed in the external service, as were most inspectors and teachers, they were hired under the Agricultural Instruction Act, which became the department's gateway to another type of activity: production and application of technoscientific agricultural knowledge. It was under similar circumstances that the department established services to support specialized production, including those for animal husbandry, the dairy industry, field crops, horticulture, apiculture, and the maple syrup industry, all created between 1915 and 1925. Technoscientific expertise flourished – the result of specialization in agronomic knowledge and agricultural production – with the hiring of employees, most of whom had been educated in the province's agriculture schools.

The external service's employees in forestry, mining, and wildlife might work in more than one sector at a time, depending on the department to which they were attached (for example, surveying and exploration performed for the forestry and mining sectors, or for development of national parks for the forestry and wildlife sectors). It therefore became more difficult to describe their specific sector of intervention, and so I chose to bring them together under the heading of "Natural Resources." For these sectors, even before scientific services were instituted, the number of staff was growing as geometers, surveyors, cartographers, and draughtspeople, responsible for sketches and cadastres, were hired. Grouped in the Department of Crown Lands after it was created under the administration of the Province of Canada, these specialists initially worked mainly in the forestry sector. The main position was land agent. The functions of these employees concerned essentially administration of public lands, and the tasks included selling, leasing, and inspecting lots, granting licences and felling permits, protecting forests, and, above all, collecting revenues.[20] However, when Premier Simon-Napoléon Parent was forced to resign following accusations of corruption related to administration of the public domain (formerly commissioner of Crown lands, he had kept this sector under his aegis once he became head of government to stimulate industrial development in the province through exploitation of natural resources and foreign investments), his successor, Lomer Gouin, decided to create a department devoted specifically to forests. In 1905, the new minister of lands and forests undertook to place administration of public lands under the symbolic authority of science by sending two young Quebecers, Gustave Piché and Avila Bédard, to study at Yale University Forest School. When they returned, Piché and Bédard established the Forest Service. The service absorbed the Woods and Forests Branch, which was collapsing under accusations of incompetence and corruption, as well as the École de foresterie, which the department annexed to Université Laval.[21] The cohorts of forest engineers trained at Université Laval made it possible for the Department of Lands and Forests to recruit qualified personnel; already, by 1914, eighteen of the twenty-seven graduates of the school since its creation were working there (see Chapter 3). In addition to responsibilities added occasionally as necessary – such as for fighting fires and insects and conservation of natural resources in the national parks and forest reserves – the main task of the Department of Lands and Forests remained surveying and inventorying, but it was also responsible for studying and developing the forest estate and its resources.

The Quebec state had few mining geologists until the late 1920s, in either the internal or the external service. With the exception of Joseph Obalski, the province's mining engineer hired in 1881, geological information about the province came mainly from the Geological Survey of Canada, founded in 1842 and performing fieldwork in Quebec since then.[22] The Bureau of Mines sometimes hired naturalists, but mainly professors from Université Laval, McGill University, or the École polytechnique de Montréal, to conduct geological explorations. It should be noted that during the first two decades of the twentieth century, the mining industry in Quebec rarely required the mobilization of people with advanced geological knowledge, in large part because the Bureau of Mines could call upon the reservoir of information that the Geological Survey of Canada had accumulated in the province. The main extractive activities were quarries supplying structural materials for the construction industry and asbestos strip mines in the Thetford region that had been in operation since the 1880s. During the 1920s, state interventions were articulated more around the scientific expertise that the Bureau of Mines acquired with a view to encouraging copper and gold extraction in the northwestern part of the province. To this end, the service sent a growing number of geologists on expeditions to explore and probe the territory (see Chapter 2).[23] In addition, it created the Division of Geology, which, in addition to recruiting its own scientists, hired academics from Quebec, Ontario, and the United States on a seasonal basis to expand exploration in the province.

Finally, in the wildlife sector, the two or three people employed by the Fisheries and Game Service at the turn of the twentieth century were naturalists who provided support to members of private clubs. On a seasonal basis, the department also hired several hundred game and fishing wardens from nearby communities, which had to be kept from pillaging Crown lands.[24] Nevertheless, the minister admitted that the public domain, despite its being almost completely privatized by An Act to Facilitate the Formation of "Fish and Game Protection Clubs" in the Province, was too extensive in area to be subject to rigorous surveillance, and so this responsibility was passed on to private clubs in exchange for exclusive hunting and fishing rights.[25] In 1915, the department began to conduct a series of fish-stocking operations in conjunction with fish hatcheries throughout Quebec, after the federal government transferred to it the pisciculture stations that it was running in the province.[26] To maintain these pisciculture activities, the Fisheries and Game Service transferred management of the stations to individuals outside of the civil service.[27] Activities aimed at improving the hunting and fishing environment took a new direction after Bertram William Taylor was hired, and in 1930 a biology service was created to support the Hatchery Service (see Chapter 4).[28] The Biological Service undertook a series of studies on freshwater fish and their environment, but it was only with the creation of the Quebec Biological Bureau in 1943, after the Department of Hunting and Fishing was established in 1939, that the Quebec public administration had a corps of scientists for the sportfishing sector (see Figure 1.1).[29] Terrestrial fauna and hunting would receive technoscientific support only much later.[30]

FIGURE 1.1 The opening of the Saint-Faustin hatchery enabled the Fisheries and Game Service to consolidate its pisciculture activities in the Laurentians and to launch research projects under the aegis of the Biological Service. | The provincial fish hatchery at Saint-Faustin in the Laurentians, by Claude Décarie, 1944. *Source:* BANQ-Vieux-Montréal, Fonds ministère de la Culture et des Communications, Office du film du Québec, Documents iconographiques, E6, S7, SS1, D18273-1828.

The expansion of the Quebec public administration and the recruitment of qualified personnel in the agricultural, forestry, mining, and wildlife sectors resulted from several crucial factors: the establishment of applied-science programs in universities, the multiplication of responsibilities of the Quebec government in view of industrialization and occupation of the territory, and the technoscientific infrastructure of the federal state. The increase in scientific personnel – especially university graduates – in the public administration was an indication, first of all, of changes in the educational opportunities offered in the province. A number of institutions of higher learning had existed in Quebec for several decades, but the addition and maintenance of forestry and agriculture programs was a much more recent phenomenon (McGill had offered courses in agriculture in the mid-nineteenth century, but the program had not been a success).[31] The situation was a bit different in the mining sector, which was without a specific site for training before the creation of the École des mines attached to the École supérieure de chimie at Université Laval in 1938.[32] Although training mining engineers in a French-language institution became a priority for the Union nationale government in the 1930s in the wake of mining development in northwestern Quebec, the École polytechnique de Montréal – which was bidding to host the new school – and McGill University had been acting as centres for training in this sector for several decades. In fact, a number of graduates from the engineering and geology programs at the École polytechnique and McGill joined the provincial administration – some after having worked at the Geological Survey of Canada, others after having participated as students in geological explorations that the service funded by granting contracts to professors in these universities (see Chapter 2). Young people wishing to acquire training in the mining sector also attended the mining school at Queen's University in Ontario – to which the Quebec government sent scholarship students starting in 1934 – before finding employment with the Geological Survey of Canada.[33]

For the scientific services that were growing more slowly, the Quebec public administration compensated for a lack of personnel by turning to external resources. For the state, the presence of these people in the field also explained the lack of urgency to modify the provincial technoscientific organization and provide it with qualified staff.

On the one hand, departments were calling upon farmers' associations or fish and game clubs that were responsible for gathering and distributing information on the availability and condition of natural resources and on exploitation techniques. For instance, the Department of Agriculture (attached first to Public Works, then to Colonization) funded a series of activities undertaken by county agricultural societies, horticultural associations, fruit growers' associations, and breeders' associations.[34] The activities included organizing fairs and exhibitions, holding contests and demonstrations, inviting speakers, importing purebred animals, administering breeding records, and producing and distributing journals and pamphlets.[35] Because funding was conditional on submission of a report on the groups' activities, the department could generally count on their cooperation; its own staff worked mainly on coordinating all of the groups' reports for production of the department's annual report. Conservation of wildlife resources was also handed over to private associations, the fish and game clubs, which were entrusted with managing and improving large estates.[36] The clubs, however, proved to be less cooperative when it came time to send in their reports, whether to avoid publicizing the wealth of their estates or for fear of having the cost of leasing rise. As they did not depend on state funding, unlike the farmers' associations, the superintendent of fisheries and game had to come up with other strategies for obtaining information on wildlife populations.

On the other hand, the Quebec public administration left things to the federal scientific institutions, such as the Geological Survey of Canada and the Experimental Farms Branch of the Department of Agriculture in Ottawa. The federal intrusion into spheres of provincial jurisdiction (natural resources) or shared jurisdiction (agriculture) took different forms, stemming in the former case from the seniority of the Geological Survey of Canada, on which the provincial administration had come to depend to some extent for evaluations of the mining potential in the territory, and in the latter case from a trend toward centralization of agronomic research justified, in retrospect, by invoking provincial constitutional responsibilities in education.[37] Although it had better financial and human resources to perform experimental work, the federal Department of Agriculture left it up to the provinces to disseminate the knowledge produced by its civil servants.

In this regard, Quebec was not an exception among Canadian provinces. Indeed, the technoscientific infrastructures of the federal Department of Agriculture, starting with its network of experimental farms operated by the central farm in Ottawa, seem to have played a similar role in the other provinces. Because universities and their affiliated schools of agriculture had been created belatedly (with the exception of Ontario and Nova Scotia, where agricultural colleges dated from 1874 and 1885, respectively), the provinces' capacity to establish agronomy services was limited, and most provincial administrations turned toward extension and education activities.[38] Even in Quebec and Ontario, it was not until the 1910s that provincial administrations integrated technoscientific competencies and became more interventionist, rather than basing their approach on the delegation of responsibilities to associations.[39] In contrast, the Geological Survey of Canada had a stronger presence in Quebec than in Ontario and British Columbia, whose provincial governments were quick to encourage the development of a bureau of mines by hiring qualified personnel and funding multiple geological explorations as the mining industry began to prosper in their territories in the late nineteenth century. In the forestry sector, the federal administration had little presence in Quebec, unlike in New Brunswick, Nova Scotia, and British Columbia, where the forest inventories conducted by the federal Commission of Conservation during the 1910s breathed some energy into the instigation of technoscientific work in provincial forestry services.

Finally, it should be noted that the Quebec state was able to recoup knowledge and skills acquired by federal agencies when it began to consolidate its scientific organization in agriculture and geology. Among other things, the provincial agricultural department recruited from the federal civil service the directors of the Laboratoire d'analyse officielle de la province de Québec, which it had founded in 1888; Léon Anatole Tourchot, analyst in the customs laboratory of the federal Department of the Interior, became director of the provincial laboratory in 1903, and A.T. Charron, assistant to the Chemist of the Dominion in the Department of Agriculture, succeeded Tourchot in 1915, the year when he received a PhD in chemistry from the University of Ottawa. In the mining sector, the Geological Survey of Canada became a recruitment pool for the Quebec Bureau of Mines; Obalski's successor, Théophile-Constant Denis, came from the Survey, as did John-A. Dresser, appointed director of the Division of Geology when it was created in 1929.

DIVERSIFICATION OF ACTIVITIES BY THE SCIENTIFIC SERVICES

With both greater means at its disposal and a growing list of responsibilities, the administrative apparatus of the province of Quebec hired technicians and scientists and assumed increasingly specific tasks in the early decades of the twentieth century. The presence of qualified personnel within the public administration allowed, in turn, for diversification of activities by the scientific services. To characterize the evolution of these technoscientific activities, I have identified six types of activities upon which expenditures were made in the different departments, grouped under three headings: the *educational, descriptive,* and *experimental enterprises* (see Table 1.3). Through an examination of the public accounts, I have established, for each five-year period, an average proportion of expenditures devoted annually to each activity for the "Agriculture" and "Natural Resources" fields, as defined above (see Appendix).

TABLE 1.3 Technoscientific interventions

Enterprise	Activities
Education	Training (teaching, chairs, scholarships)
	Dissemination (lectures, demonstrations, journals)
Description	Measurement (analysis, laboratories, surveying, soil classification, statistics, mapping)
	Inventorying (collections, exploration, sampling, descriptions)
Experimentation	Studies (investigations, research, experiments)
	Applications (nurseries, pisciculture, inspection, surveillance)

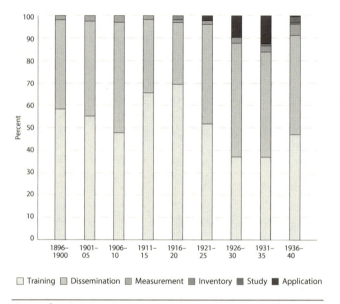

FIGURE 1.2 Technoscientific interventions and exploitation of agricultural resources

Source: *État des comptes publics de la province de Québec, 1896–1940*.

Figures 1.2 and 1.3 present the general outlines of the evolution in relative importance of technoscientific activities in each field. Not only did certain enterprises predominate in each field, but the relative importance of each activity followed a distinct pattern related to the centrality of the respective resources at play in the provincial economy and to the development of scientific institutions. In "Agriculture," the growing role of activities related to descriptive and experimental enterprises did not draw impetus away from the educational enterprise – training of technoscientific personnel and dissemination to farmers. In "Natural Resources," measurement and surveying were the main intervention tools, whereas the experimental enterprise gained in importance and education was always in a negligible position.

The educational enterprise was at the core of technoscientific activities in agriculture – even more so once the Department of Agriculture stopped relying on agricultural societies to promote improvement of agricultural production.[40] Under the leadership of Édouard-André Barnard, inspector of agricultural societies from 1872 to 1876, then director of agriculture from 1877 to 1889, the department increased dissemination activities through agricultural speakers and, especially, the publication and distribution of the *Journal d'agriculture illustré*.[41] In both cases, the effectiveness – and the very existence – of

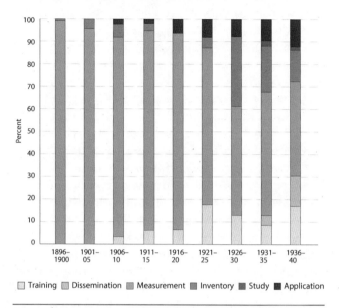

FIGURE 1.3 Technoscientific interventions and exploitation of forestry, mining, and wildlife resources

Source: *État des comptes publics de la province de Québec, 1896–1940*.

the vehicle depended on smooth operation of the farmers' clubs. In existence since the last quarter of the nineteenth century, the speaker program and the journal became the main springboards for the Department of Agriculture's educational enterprise. The department complemented its dissemination efforts with training activities by funding agriculture schools at different levels and by creating its own schools. For instance, it provided financial support to the Société d'industrie laitière de Québec to create the École de l'industrie laitière in Saint-Hyacinthe, in order to have a training site for its factory inspectors and instructors who visited butter and cheese factories; at the same time, the agricultural middle schools trained the younger generation of farmers.[42] Finally, the department benefited from the existence of agricultural schools at the university level, which it funded in whole or in part to train the inspectors, veterinarians, and agronomists who joined its ranks to provide it with a means of action and a voice throughout the province.

In the late nineteenth century, the department found new ways to disseminate agronomic knowledge. First, it provided financial assistance to farmers whom its agents supervised in the management of their facilities in order to promote "good practices to the surrounding farming population," or to introduce new crops or breeds. Second, it operated its own demonstration sites – farms, orchards, and fields – which, under certain conditions, became sites for experimentation and specialization of agricultural production. In addition to arboricultural stations, demonstration orchards, and poultry stations (discussed in detail in Chapter 5), fourteen clover and corn experimental stations were established between 1912 and 1920 to provide information about local growing conditions and provide demonstration sites for "horticultural industrialization."[43]

Training at the university agricultural schools paved the way for another type of technoscientific endeavour. Even though their mandate mainly fit within the educational enterprise, the department's official agronomists – graduates in agricultural science – occasionally performed descriptive and experimental activities. It must be noted that an informal division of scientific labour was established between the two levels of governments in Canada: stations of the Experimental Farms Branch under the federal Department of Agriculture conducted agronomic research, and the Quebec Department of Agriculture saw to disseminating the results of the research. In the late 1920s, however, the Quebec department began to expand its measurement and inventorying activities, previously limited to the Laboratoire d'analyse officielle de la province de Québec, which was on the same site as the École de l'industrie laitière in Saint-Hyacinthe. Up to then, it was in this laboratory and in the experimental fields attached to the dairy school that analyses were conducted, mainly to assess soil fertility or milk quality. But in 1929, as part of a reform aimed at restoration of Quebec agriculture established by minister Joseph-Léonide Perron, the department began a research program to encourage regionalization of agricultural production (see Chapter 5).[44] Achieving this objective required that a map of regional agricultural potential be drawn so that production could be concentrated according to the agricultural and pedological characteristics of each region.[45] Economic and agronomic studies were conducted to determine regional production that would be integrated into the mixed polyculture system that sustained the dairy industry in the province. These investigations were based on an expansion of the work done at the Laboratoire d'analyse officielle after the department moved its infrastructure to Quebec City in 1933 to orient its work toward analysis and "methodical mapping of the soil of the province by region."[46] In spring 1937, the Plant Protection Service – which also had entomology and botanical laboratories in Quebec City for identification of harmful insect species and plant diseases – created a network of field laboratories in Granby, Saint-Hilaire, Saint-Martin, and Île d'Orléans to study agricultural pests (see Figure 1.4).[47] Aside from these investigations, the experimental enterprise of the Department of Agriculture also comprised activities aimed at applying knowledge on the cultivation and distribution of seedlings to support the

FIGURE 1.4 In the late 1930s, as the Department of Agriculture was centralizing its research activities in laboratories in Quebec City, some regionalization occurred with the establishment of field laboratories in various agricultural regions. | Field laboratory in Granby, by Paul Carpentier, 1938.
Source: BANQ-Québec, Fonds ministère de la Culture et des Communications, Office du film du Québec, Documents iconographiques, E6, S7, SS1, P10489.

development of the fruit industry. The department began to grow seedlings to establish demonstration orchards and supply arboricultural stations in 1907, when the Department of Lands and Forests established a tree nursery in Berthierville, between Montreal and Trois-Rivières. Ten years later, it created its own nursery in Deschambault, near Quebec City, where it established testing grounds to experiment with growing rustic plants, propagating endangered varieties, and acclimatizing plant and animal species.[48]

The presence of a large number of employees with bachelor's degrees in agricultural science, the specialization of agronomic knowledge, and the desire to concentrate specific agricultural productions in certain regions engaged the department in a *descriptive enterprise,* involving much measurement and inventorying, to provide data for investigations and applications. The *experimental enterprise* was developed from a network of demonstration sites, stations, and laboratories, and from the creation and expansion of the Deschambault tree nursery and the Conseil provincial des semences, which conducted comparative tests aiming to improve seeds and the propagation of remunerative field crops. The *educational enterprise* nevertheless remained at the heart of the department's operations.

Conversely, for forestry and mining resources, measurement and inventorying, mainly through surveying and geological exploration, were the main technoscientific activities. The few educational initiatives required little money. Beyond partial funding of the École de foresterie and the École d'arpentage (School of Surveying) starting in 1910 and of the École supérieure des mines starting in 1938, and the creation of the Paper-Making School in Trois-Rivières and a school for forest rangers in Berthierville in 1923, training efforts in "Natural Resources" consisted of granting scholarships.[49] Outside of educational institutions, dissemination activities were limited to lectures on forest protection and conservation and courses offered occasionally to mining prospectors.

In "Natural Resources," technoscientific interventions mainly revolved around the *descriptive enterprise*. The Survey Branch, in existence since 1866, conducted the Department of Crown Lands' basic activity. In addition to measuring and marking the boundaries of public lands, the department sought to identify and make visible the riches that these lands contained in order to encourage occupation of the territory and exploitation of the resources. For example, it sent explorers to survey the land after acquisition of territories in the northwestern part of the province in 1898. Exploration reports included an assessment of ores, hydraulic energy, and wildlife resources to attract settlers and industrialists to the new areas.

Once the public lands were explored and surveyed throughout the province, surveying activities embraced new goals, as the state set out to define land-occupation schemes. In 1905, the government created the Department of Lands and Forests to undertake classification of soils and division of the public domain between land for agriculture and land for forestry. These undertakings followed the recommendations by the Colonization Commission, which had tabled its report the previous year. After lots suitable for colonization were transferred to the Department of Colonization, Mines, and Fisheries in 1921, the Forest Service was relieved of administration of lots but remained responsible for classification of townships and exploration of the territory.[50] The Survey Branch, meanwhile, continued with allotment of land, definition of timber limits, and development of the reserves and parks being created by the Department of Lands and Forests.

Measurement remained a core activity of the department. With forest engineers trained at the École de foresterie at Université Laval now in its employ, the Forest Service could proceed with inventorying tree resources, preparing statistics on wood production and trade, and conducting silviculture studies.[51] In addition to taking on responsibilities occasionally added as the situation warranted, along with recruiting and training qualified personnel, the Forest Service created opportunities to undertake descriptive activities by having a statute on forest inventories adopted so that it could conduct studies on the regeneration of trees and on the availability of forest products as a function of the species present in the limits allotted to forestry companies.[52] Forest inventories and studies on the regeneration cycle served as the basis for formulating working and felling plans. In 1938, sixteen years after adoption of the Loi sur les inventaires forestiers, 60 percent of the leased forest had been inventoried, and almost three-quarters (73.8 percent) of the inventoried forest had been included in a silviculture plan.[53]

In parallel with this diversification of descriptive activities, the Forest Service undertook to improve the forest estate by protecting the forest against fires and insects and setting up a network of tree nurseries around the Pépinière provinciale de Berthierville (Provincial Forest Nursery Station at Berthierville) for production of a "pulp forest."[54] During the 1930s, research within the department was expanded with the creation of the Commission des produits forestiers to conduct economic and statistical studies on the use of wood, and then of a research station at the provincial tree nursery.[55] The result of a collaboration between two levels of government, this "first forestry research station" was composed of laboratories set up in new buildings at the Forest Nursery Station at Berthierville.[56] For the provincial minister, the idea was to "avoid duplications" and recognize "the fact that plant diseases and insect infestations ignore borders."[57] At the research station, entomologists such as Lionel Daviault, of the federal entomology service, conducted studies on the propagation of harmful insects, such as the hemlock looper in the Côte-Nord region and the spruce sawfly on the Gaspé Peninsula. At the same time, forest pathologists, such as René Pomerleau of the provincial Forest Service, looked into the damping-off of coniferous seedlings.[58] In 1935, the Forest Service moved the Écoles des gardes forestiers (Rangers School) to Duchesnay, where it also conducted research on the flammability index of timber stands (see Figure 1.5), and the following year it established the entomology and meteorology bureaus in Berthierville.

In the mining sector, technoscientific activities revolved mainly around measurement and surveying throughout the period under study. For mineralogical and chemical analysis of ore samples submitted by prospectors, the Bureau of Mines established the Provincial Government Assay Laboratory at the École polytechnique de Montréal in 1911. Previously, it had sent this work to a New York company and then to a chemist in Montreal, Milton L. Hersey, whose engineering consulting firm specialized in mineralogical identification.[59] The gold-mining industry was developing in the northwestern part of the province during the 1920s, and in 1924 the bureau subsidized private laboratories in Abitibi to accelerate analysis of samples, at the same time that Thetford Mines became host to a public laboratory to assess the

FIGURE 1.5 Initially oriented toward training forest engineers and acclimation tests, the Forest Nursery Station at Berthierville came to host various research activities directed toward protection of trees against insects, diseases, and fires. | Berthierville nursery: rain gauge, by T. Deslauriers, 1942. *Source:* BANQ-Québec, Fonds ministère de la Culture et des Communications, Office du film du Québec, Documents iconographiques, E6, S7, SS1, P6762.

FIGURE 1.6 In the 1930s, the Bureau of Mines gradually stopped performing analyses in the mining regions, opting instead to concentrate on laboratory work in Quebec City in a complex created in 1936. | Laboratory (Department of Mines), by Gérard Bastien, 1938. *Source:* BANQ-Québec, Fonds ministère de la Culture et des Communications, Office du film du Québec, Documents iconographiques, E6, S7, SS1, P17277.

quality of asbestos fibre and standardize its classification for marketing purposes. Then, in 1933, the Department of Mines centralized the ore-analysis activities into its premises by creating a laboratory in Quebec City. This relocation and expansion (see Figure 1.6) was intended to serve the needs of the department's technical personnel; its industrial clientele was referred to the laboratory at the École polytechnique.[60] As for surveying activities, the number of explorations led by the Division of Geology grew rapidly starting in 1929, after the recruitment of university-educated geologists with experience in a state agency or a private company.[61] The extension of this work and the reorganization of the laboratories signalled a specialization of knowledge with regard to mining exploration and exploitation. The department reacted to these developments in February 1939 by creating two units to replace the Division of Geology: the Division of Geological Surveys and the Division of Mineral Deposits.

Finally, a similar phenomenon took place in the fish and game sector. The start-up of research projects was based on extension of the infrastructure underlying practical applications. First defined in relation to the needs of the Hatchery Service, technoscientific activities embraced hydrobiological research in the 1930s to intensify fish stocking in watercourses to be used for sportfishing.[62] Here, the initiation of research programs was a result of the infrastructure of the federal administration, rather than the availability of qualified personnel, which had stimulated such programs in agriculture and geology. Indeed, it was when the federal minister of fisheries transferred its hatcheries to the provincial Department of Colonization, Mines, and Fisheries in 1915 that the latter created a hatchery service, first to administer these facilities, then to organize and improve fish-stocking activities.

Although the availability of qualified personnel encouraged the diversification of technoscientific activities in the administrative apparatus, most interventions by each department concerned a specific enterprise throughout the period studied. This enterprise continued to be strongly influenced by historical conditions that led different governments to engage in regulation of the exploitation of natural resources and occupation of the territory.

The significance of the educational enterprise observed in the field of agriculture was not unrelated to the institutions and practices inherited from the Province of Canada, in both Quebec and Ontario. Technoscientific activities involved in administration of the public domain, and more particularly of the forest, were different in nature – even though, here too the personnel and organization of the commissioner of Crown lands of the Province of Canada were transferred to provincial administrations immediately following Confederation. In both Quebec and Ontario, the respective departments of agriculture were responsible mainly for funding local societies and sector-based associations for the education of farmers and the marketing of their products. Both provinces also financed the creation of university-level agricultural schools; this was a question not simply of upgrading the education of farmers' sons but also of training scientists and technicians to meet the needs of increasingly specialized and market-integrated agricultural production. In the early twentieth century, when provincial administrations were managing to set up a few laboratories, the federal infrastructure, in place for several decades, was in a clearly advantageous position when it came to recruiting scientific personnel and launching large-scale research programs.[63]

In forestry, on the other hand, the growth in provincial organizations saw different trajectories with regard to both personnel numbers and techno-scientific activities. It is true that in Ontario, the professionals were crossing swords with an administration that was relatively indifferent to their advice during the early decades of the twentieth century; even though they had little voice, many foresters joined the public service, in which, despite little room for maneuver compared with Quebec or even British Columbia, they were responsible for reforestation and protection of the forest.[64] Similarly, Crown lands represented, after the federal government subsidies, the second-largest source of revenue for the public treasuries of Quebec and Ontario. In Ontario, the Department of Crown Lands was thus perceived as "*the revenue* department," as political scientist John Hodgetts put it.[65] After the Great War, the relative importance of this revenue stream fell slightly in favour of new fiscal initiatives, although, during the interwar period, Crown Lands remained a privileged ground for intervention, as attested to by budgeted and realized expenditures – in Quebec, at least.[66] It was during this period that the scientific services related to exploitation of natural resources saw both quantitative and qualitative enhancements.

As fiduciaries of Crown lands, the provinces saw to protecting this public asset, for which they retained responsibility and which was to serve the general interest. It was different for agriculture, in which settlers became owners of the land that the state had allotted to them when it was recognized that they had fulfilled the conditions for clearing and cultivating the land and building a house. On the colonization front, where an individualistic pioneering spirit dominated, state interventionism ended, as the state ceded its rights over the public domain.[67] To regulate exploitation of agricultural resources, the state had to act indirectly, through education of farmers or through associative forces.

In these examples and others cited above, the experience of Quebec seems to have been similar to that of Canada as a whole, although a few specificities emerge. Both the other provincial administrations and departments of the federal administration hired scientists who had graduated from North

American and European universities. However, English-speaking students enjoyed greater mobility for studying and working in institutions outside their provinces. In British Columbia, for example, despite the creation of the provincial university in 1908, and of its forestry faculty some years later, in 1920, the public service recruited foresters from universities in Washington and Oregon, on the American west coast, and from universities in eastern Canada. A telling example of this mobility is that of the first forester hired in New Brunswick: a graduate of the University of New Brunswick, which had had a forestry school since 1907, G.H. Prince began his career in the British Columbia forestry service in 1912, before his own province created its service with the Forest Act of 1918.[68] Of the forest engineers trained at the University of Toronto, some joined the Bureau of Forestry of the Ontario Department of Land, Forests, and Mines, but the slow start-up of this service, in terms of both activities and personnel, led a number of them to find work in the Canadian Forest Service or another provincial administration.[69] The creation of institutions for teaching applied sciences in universities in Quebec took on a strategic character that departments would prioritize to provide the public administration with qualified personnel.

Another difference involved the conditions in the different provinces for setting up technoscientific services, the great majority of which appeared following commissions of inquiry. In Ontario and British Columbia, the forests – and in Ontario, mining and wildlife as well – were the subject of commissions of inquiry, and the recommendations given in the reports led the provincial governments to adopt laws governing the exploitation of natural resources and to set up scientific organizations to apply them.[70] Often created following scandals or natural disasters, these commissions were intended to collate information on the availability of resources, but they also enabled governments to bring together North American experts to take stock of issues affecting the exploitation and conservation of natural resources.[71] In Quebec, the Colonization Commission preceded a series of organizational reforms for exploitation of Crown lands, although there are no solid indications that it encouraged the Liberal government to send Gustave Piché and Avila Bédard to study at Yale University with an eye to creating a forestry service. In this sector, as elsewhere, the creation of the different services seems generally to have been the result of political expediency.

THROUGH THIS SYNCHRONIC and diachronic assessment of the relative importance of the educational, descriptive, and experimental enterprises within the public administration, I have uncovered certain key elements of the technoscientific interventions by the Quebec state from the late nineteenth century to 1939. As the provincial government funded or took the initiative to create its own educational institutions to train qualified personnel, the implementation of scientific services was accompanied by the internalization of skills through the recruitment of a corps of technicians and scientists. The efforts devoted to hiring qualified personnel resulted in the diversification of interventions and increased the complexity of organizations simultaneous with specialization of technoscientific knowledge. These phenomena, which attest to the relative autonomy not only of the public administration but also of scientific institutions, contributed to the modernization of the mechanisms for state intervention in Quebec.

The next four chapters provide illustrations of how, thanks to educational, descriptive, and experimental enterprises, the scientists in the employ of the Quebec state formed new relations with the environment based on the defining terms for exploiting natural resources, defining territory, and shaping the landscape. As we shall see, by setting conditions for appropriating and occupying the territory, technoscientific activity extended and consolidated the state's power in the "government of men and things."

2

The Invention of a Mining Space: Geological Exploration and Mineralogical Knowledge

THE GEOLOGICAL SURVEY of Canada was created in 1842 to study geological formations and assess the potential of mineral deposits in the Province of Canada.[1] This work extended beyond both the borders of the colony and the technical aims of the Survey when the discovery of coal in the adjacent British colonies fostered a confederative outlook.[2] Following Confederation, Survey personnel were further diverted from their initial field of investigation after the acquisition of Rupert's Land and the North-Western Territory in 1870 and, a year later, the entry of a new province, British Columbia. During this period, industrialists and politicians also accused Survey personnel of generating knowledge of little economic interest and practical value.[3]

These departures from the original mandate of the Geological Survey concerned the Quebec government, which deemed it necessary to have information on the Crown lands for which it was becoming the sole fiduciary. This information would provide a basis for establishing its authority over that of adjacent provinces and private interests. It would also be useful in attracting investors, who might acquire mining claims or pay royalties for resources extracted from Crown lands, generating revenues for the provincial treasury. The geological surveys would be a powerful tool for colonization, as they described resources in remote areas likely to appeal to both frontier settlers and industrialists prepared to invest the capital necessary for extraction and processing of mineral resources. By encouraging the establishment of a mining industry in the province, geological information would indirectly create employment opportunities for a local population otherwise tempted by emigration to the United States. It was particularly important to construct the colonization discourse on these new foundations because the provincial government wanted to populate marginal zones north of the St. Lawrence Valley, where the Precambrian rock of the Canadian Shield offered relatively poor soil of low fertility, unsuitable for farming, which was usually promoted in colonization projects.[4] Finally, the Quebec mining industry was not making a significant contribution to expanding the provincial tax base, and the areas where mines were being operated, mainly in the St. Lawrence lowlands close to older settlements, had not been swept up in the new industrialism. This contrasted with the situation in other Canadian provinces, where capitalist syndicates were engaging in intense mining activity, thereby enriching public treasuries through the issuance of permits, the granting of mining claims, and the collection of royalties on ore extracted.

In this chapter, I analyze the terms and conditions under which the province of Quebec created legal and technoscientific instruments for building its geological expertise to govern the exploitation of mineral resources. The role played by the Geological Survey of Canada in the early reconnaissance of the province's mineral potential shows that the Quebec public administration initially lacked the expertise to direct resource extraction and colonization in remote areas. For the Quebec government, it was essential to make visible the presence of minerals in the province's subsoil and encourage the establishment of a mining industry, especially in the part of the Canadian Shield that the federal government had transferred to the province. As mining operations in northeastern Ontario and the completion of a rail line facilitated an initial influx of prospectors and explorers to northwestern Quebec, survey and measurement activities by provincial geologists in that area provided further impetus. Both by their deployment in the territory and by their cartographic production, the geologists communicated information on a mining space appropriate for the consolidation of extractive activities in the northwestern part of the province.

Creating Visibility for an Industry and Its Space

For almost a quarter-century, the Geological Survey of Canada devoted its resources to producing a geological portrait of the Province of Canada. After the British North America Act was adopted, the Survey's exploration and surveying activities expanded to include the new provinces of Nova Scotia and New Brunswick. Once ownership of Crown lands and their natural resources was formally transferred to provincial administrations, however, the federal agency no longer saw great urgency in discovering or exploiting ores in the public domains of the older provinces. Instead, Canadian expansionism led the Survey to commit its personnel to more ambitious expeditions in the West. Ottawa hoped to stimulate Euro-Canadian settlement, and so the Survey focused its activities more specifically on describing access routes and natural resources in the recently acquired territories.[5]

In Quebec, the Survey hired temporary staff to locate ore deposits in the Appalachians region, where most of the province's mining industry was based. Aside from the Chaudière River valley, where the discovery of gold-bearing alluvial deposits drew prospectors in 1847, the main mining zones were the Eastern Townships, where extraction of copper and sulfur began in 1859, and Coleraine and Thetford townships, where asbestos-mining operations started up in 1878. To these were added several phosphate-of-lime and mica mines in the Ottawa Valley and quarries for construction materials situated, for the most part, near urban areas.[6] The Survey performed some work around these sites, but it was involved primarily in expeditions to the northern territories of Rupert's Land, beyond the borders of the provinces of Quebec and Ontario and the colony of Newfoundland.[7]

The Survey personnel deployed in Quebec were now fulfilling the organization's mission and the objectives set out by its successive directors, which responded to the demand for research of interest to Canadian industry. This was a change from the Survey's early years, when, in the absence of strong mining activity in the country, its geologists had the leisure of studying rock formations to understand the geological structure of Canada and its directors fostered the hope of building an internationally renowned scientific institution on a par with its peers in Great Britain and the United States.[8] The Survey's explorers were also asked to gather the data needed for cartographic production, a step that was considered necessary to strengthening geological knowledge about the territory. Most of this work was of limited interest to mining companies, which would have preferred having information that would help them locate minerals and assess their quality and availability.[9] In 1884, during hearings of a special committee of the House of Commons on the Geological Survey of Canada, industrialists, bankers, and prospectors demanded smaller-scale, more localized studies to provide guidance for mining operations. In its own defence, the Survey stated that it would be better able to define the mineral deposits discovered once it had built a knowledge base.

It is worth taking a closer look at the presentation on unexplored territories given by the associate director of the Survey, George Mercer Dawson, to the Ottawa Field-Naturalists' Club in March 1890.[10] First, Dawson emphasized the usefulness of the information generated by Survey personnel to obtain support from the public and government for continued exploration and mapping of Canadian regions, and he noted that trappers, lumberjacks, farmers, and then railway builders would follow in the footsteps of the surveyors.[11] He then justified the need to conduct explorations in order to gain geographic knowledge and to discover and identify forest resources and ore deposits:

> It is, therefore, rather from the point of view of practical utility than from any other, that an appeal must be made to the public or to the government for the further extension of explorations, and my main purpose in addressing you tonight is to make such an appeal, and to show cause, if possible, for the exploration of such considerable portions of Canada as still remain almost or altogether unmapped.[12]

In all of the unexplored regions that were shown on the map accompanying the presentation (see Figure 2.1), potential mineral wealth offered an incentive for proceeding with exploration, although Dawson observed that "negative information" – if, for example, no ore was found in a region that was explored – would also be useful. Even if explorers had to report that a territory had no economic value, Dawson said, their notes would "contain scientific observations on geology, botany, climatology, and similar subjects" that would justify the expense incurred for their expeditions.

Dawson's presentation is revealing in two ways. First, regarding the delicate balance between usefulness and curiosity in the Survey's work, as a geologist he clearly showed his preference for the latter, even though in his introduction he made sure to mention the practical consequences of geological explorations. Second, and partly as a corollary, the lands slated for future expeditions were outside of Quebec – south and east of James Bay (region no. 15) and in the Labrador Peninsula (region no. 16) – a vast area that Dawson called the "northeastern territory" and that the federal government was readying to transfer to the province. However, it seemed unlikely that prioritizing exploration of these territories would garner information likely to stimulate Quebec's mining industry. The provincial government wanted to know about its geology in order to publicize ore deposits and encourage the growth of a mining industry, but the Survey's contribution to this goal seemed unlikely in the near future, especially after Dawson succeeded Alfred Richard Cecil Selwyn as director in 1895.

Dawson directed the Survey's activities around the program that he had presented to the Ottawa Field-Naturalists' Club. He sent most of his personnel to the "great unexplored territories" of the Canadian West and the coasts of British Columbia, leaving just a few geologists behind to explore and survey the territories north of Quebec's borders. Apart from these activities, the Survey limited its work in the province to analyzing the mineral

The Invention of a Mining Space 47

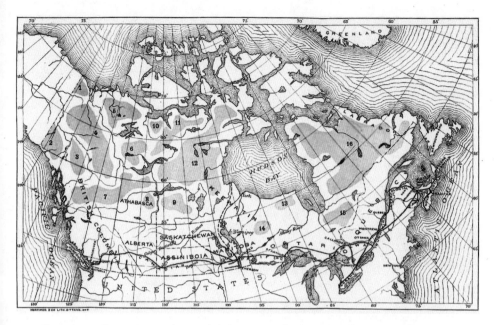

FIGURE 2.1 George Dawson's unexplored territories. | *Source:* George Dawson, "On Some of the Larger Unexplored Regions of Canada," *Ottawa Naturalist* 4 (1890): 36.

formations around existing industrial sites. Aside from sending its own geologist, Robert Chalmers, to study the geology of the Chaudière River to stimulate prospecting of alluvial gold in the Beauce region, the Survey mandated academics for this purpose. A professor of mineralogy and geology in the faculty of arts at Université Laval, J.C.K. Laflamme, travelled through Champlain, Montmorency, Charlevoix, and Saguenay counties, on the north shore of the St. Lawrence River, to study Laurentian formations between 1883 and 1892, and John A. Dresser, of St. Francis College and McGill University, surveyed the copper deposits and the serpentine belt – with its veins of asbestos and chromium – in the Eastern Townships between 1898 and 1910.[13] In the latter region, where a flourishing copper-mining industry was regarded by a contemporary observer as among the most prosperous worldwide, the Survey combined scientific and practical interests to produce detailed maps.[14]

The Survey was therefore not entirely oblivious to the practical considerations behind its creation, but these did not determine the activities that it carried out. These considerations were also addressed by the Mines Branch of the federal Department of the Interior, to which the Survey had been attached since the 1884 hearings. The branch was in charge of territories outside of provincial boundaries and mining statistics for the country as a whole.

In reaction to the mandate of this newcomer, the Survey began to publish an annual "statistical report on production, value, exports, and imports of minerals in Canada." Furthermore, its personnel stepped away from systematic mapping and description to focus on comprehension of geological structures and rock formations in Canada.[15]

During the same period, provincial administrations also established their own bureaus to stimulate mining in their respective territories.[16] With extraction of metallic ores in the Sudbury region expanding quickly, the Ontario Bureau of Mines, created following a commission of inquiry on the mining industry, took charge of mineralogical assay and technical assistance to companies in the province in 1891.[17] A succession of gold rushes and the start of coal extraction on Vancouver Island spurred the British Columbia government to institute a bureau of mines in 1895 to consolidate its personnel – inspectors and commissioners responsible for registering claims and issuing licences – to which it added a mineralogist and an assayer.[18] Before the end of the nineteenth century, the bureaus of mines in Ontario and British Columbia hired geologists to conduct geological explorations and mineral reconnaissance.

In Quebec, where the Bureau of Mines had been formed a few years earlier, similar objectives benefited from neither equivalent institutional resources nor as powerful an industry. The concentration of mining interests around copper and asbestos in the southern part of the province may explain why the bureau was slow to investigate whether there were ore deposits in the area contiguous to that which was making northeastern Ontario wealthy at the turn of the century.[19] Because, among other things, the main thrust of the Survey's efforts – spatially and scientifically – did not correspond to the economic concerns of the province, the Quebec government created tools to direct mining efforts and conquer the subsoil resources within its borders.

Although its main impact was on the mine ownership system, the General Mining Act, adopted in 1880 by the Quebec government, also authorized the conducting of "geological surveys or other searches in order to ascertain what lands contain ores or valuable deposits."[20] Previous legislation in the mining sector had concerned only gold, silver, and phosphate. As new operations were built around asbestos extraction, the adoption of this framework law broadened the range of governmental action and encouraged production and dissemination of knowledge specific to the mining industry. Indeed, the General Mining Act "defined the terms for occupation and acquisition of the territory for mining purposes," putting an end to the regime that lumped mining rights and surface rights together and instituting the principle that each could be sold separately.[21] As the purpose of the statute was to make the province's mines productive, it was important that their mineral potential be determined and assessed.

The statute conferred ownership of the subsoil on the Crown, as the "only solid basis of intervention for regulatory measures and for collection of revenues," in order to stimulate mining operations.[22] This provision gave the province the power to ensure that mines would be operated, and not left unexploited, so that rights were not granted to speculators who wished to obtain land with valuable subsoil but did not have the skills to appraise the mineral riches on their property or the capital necessary to exploit them.[23] Individuals were obliged to exploit mineral assets as necessary to improve the territory and keep speculation from impeding expansion of the mining sector.[24] The permit system imposed the requirement to prospect a certain area of land within a prescribed period, perform a determined amount of work on the staked territory, and operate claims on a continual basis or risk having the mining rights confiscated.

Ownership of mines and the terms for their operation were two aspects of the government of mineral resources set out in the General Mining Act of Quebec. Knowledge of the wealth underground and incentives to assess and extract it were the statute's main leverage. Because the Crown was the owner of ore deposits situated in territories that had not been surveyed, the Quebec government had every reason to publicize its subsoil riches.[25]

As the General Mining Act underlined the legal relevance of knowledge of the subsoil by separating ownership of the surface and of what was underground, Premier Joseph-Adolphe Chapleau was looking to strengthen certain provisions of the statute when he hired a mining engineer with a degree from the École des mines de Paris, Joseph Obalski, in 1881.[26] The same year, the Geological Survey of Canada moved its head office from Montreal to Ottawa to fulfill its transcontinental mandate and reflect its recent integration with the federal bureaucracy.[27] In Quebec City, the choice of Obalski expressed, first and foremost, the desire to assess and transform mineral resources rather than simply to generate geological knowledge of the provincial territory.[28] Indeed, the training dispensed at the École des mines de Paris revolved mainly around mineralogy and metallurgy, and this was the knowledge that was deemed necessary, if not sufficient, to application of the General Mining Act of Quebec and justified the hiring of the French engineer.[29]

Initially attached to the Department of Railways, for which Premier Chapleau acted as minister, in 1882 Obalski moved to the Department of Crown Lands, where his first task was to encourage the exploitation of mineral resources in the province. Given the intensity and success of mining activity elsewhere in Canada, especially Ontario and British Columbia, Obalski was interested in promoting intensive exploitation of Quebec's subsoil by conducting explorations and publishing the results of the reconnaissance done and the knowledge gathered. The creation of a mining industry in the province depended on the mobilization of international capital, which was targeted with publicity initiatives (see Figure 2.2) that not only described the mineral wealth of Quebec and the possibilities for processing certain ores but also highlighted the legal provisions that encouraged ownership and operation of mines.[30]

Obalski first undertook to produce a survey of ore deposits in Quebec by compiling and synthesizing the work done by the Geological Survey of Canada since its inception.[31] Then he appended to his annual report a supplement on mining production in the province, drawing on the statements that each company was required to submit to the department under the General Mining Act.[32] For a number of years, the Survey had been publishing a report on national mineral production, but Obalski deemed it insufficient. By publishing his own supplement, he wanted to illustrate the Quebec administrative capacity to compile province-wide information and formulate policies that matched the realities of the industry. The supplement also revealed his close relationship with the industry, as companies voluntarily submitted annual statements on their activities, even though many took exception to doing so.[33] Presented alongside results of explorations and studies on metals processing, the statistics in the annual report, *Mining Operations in the Province of Quebec*, helped to provide information about the province's mineral resources to operators and prospectors and to build a national industry.

In parallel with the statistical cartography of mining operations, Obalski engaged the Bureau of Mines in production of knowledge, among other things, by mandating naturalists and academics to go on expeditions in order to assess ore deposits and probe still-unexplored territories. In the thirty years during which he supervised mining development from within the provincial administration, Obalski was largely responsible for exploration of promising sites around the areas in operation and in new territories. However, his administrative workload limited his opportunities to undertake geological expeditions himself. Usually, it was during trips to inspect mining operations that he had a chance to study phosphate and mica deposits in the Ottawa Valley, copper and chromite in the Eastern Townships, and rock formations in the Gaspé area, where oil deposits attracted his attention. Otherwise, he had to rely on explorers outside of the bureau, to whom he assigned the task of evaluating mineral assets and potential for colonization, such as naturalist Henri de Puyjalon, who surveyed the Côte-Nord region in the late nineteenth century. He also hired academics to explore newly acquired territories or old colonization lands that the Geological Survey of Canada had already surveyed and where the Bureau of Mines wished to see a mining industry emerge or

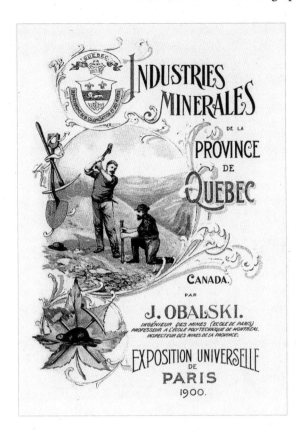

FIGURE 2.2 The superintendent of mines, Joseph Obalski, took advantage of world fairs to highlight the province's mineral riches in an attempt to attract foreign capital in order to diversify the Quebec mining industry. | Mining as a national industry. *Source:* J. Obalski, *Industries minérales de la province de Québec* (Quebec City: Dussault et Proulx, 1900).

grow. Sent to the north shore of the St. Lawrence by the Geological Survey of Canada, J.C.K. Laflamme was also mandated by the Bureau of Mines to study natural gas deposits around Louiseville.[34] Professor Émile Dulieux, director of the mining department at the École polytechnique de Montréal – with which the bureau established close connections (see below) – led expeditions to Chibougamau and the Laurentian region in 1907 and 1908.[35] Finally, two years after he arrived at the Bureau of Mines in 1906 as assistant superintendent, J.-H. Valiquette, an engineer with a degree from the École polytechnique de Montréal, led an expedition to Rivière-aux-Outardes and Manicouagan.[36]

However necessary these geological explorations were to creating a mining space, making the mineral resources visible could not in itself generate an industry on the borders of the ecumene where companies and populations would gather to expand the province's colonization area. Availability of ore, a proper road infrastructure, an increase in metal prices on stock exchanges, and technological and industrial developments in extraction and processing were all factors that affected the availability of capital to exploit an ore deposit. The role of geology and mineralogy was to pinpoint the location and content of the ore, which was only one condition for its extraction. Systematic reconnaissance and exploration still had to be conducted.

Exploration and Integration of the Middle North of Quebec

The episodes surrounding reconnaissance and exploration of the region around Lakes Mistassini and Chibougamau offer a trenchant example of the difficulties that the Quebec state was facing in taking charge of the province's geology and creating a mining space. With the prospect of having a territory added northwest of its borders (which was done in 1898, when the Parliament of Canada adopted the Quebec Boundary Extension Act, 61 Vict., c. 3), the commissioner of Crown lands sent his department's surveyors to the northern part of the province to inventory its assets, trace out travel routes, and assess the possibilities for colonization.[37] Three years later, in 1883, the Société de géographie du Québec asked that an expedition be sent to survey the "mysterious Lake Mistassini," which was thought to be so large – comparable in size to the Great Lakes – that there was talk of an inland sea.[38] The commissioner acceded to this request and negotiated joint funding of an expedition with the Geological Survey of Canada. The parties agreed that it would be headed by a veteran Quebec surveyor, John Bignell, assisted by Arthur Peter Low, a young geologist with a degree from McGill University, who had recently been hired by the Survey. Bignell was recalled to Quebec City before fulfilling his mandate, and Low became leader of the expedition, returning in 1885 to measure Lake Mistassini. A quarrel between Bignell and Low gave rise to a series of public disputes about the size of the lake. Low accused Bignell of basing himself on Indigenous legends to exaggerate the lake's area, whereas Bignell took Low to task for his alleged cupidity and careerism and claimed that he had simply reproduced what the Survey's geologists, James Richardson and Walter McOuat, had recorded in their reports during exploration missions between north of Lake St. John and James Bay in 1870 and 1871. An expedition led by a professor from the University of Toronto, William James Loudon, in 1889 corroborated Low's results and settled the controversy, though it did not end the interest that the Société de géographie de Québec had in this region north of the province's borders.[39]

This controversy revealed two spaces competing against each other to become the main access route to the northern territory: one, recently and only partially opened by the colonization movement, north of Pontiac County, and another north of two slightly older colonization regions, the backcountry of the Saint-Maurice Valley and Lake St. John. In both spaces, the main feature was a large lake that offered good potential for siting a frontier community, but the climate was so inimical – and the possibilities for farming so slim – that only the presence of natural resources essential to the Second Industrial Revolution could ensure the survival of settlers. There was, however, reason to hope that interesting ore deposits would be found, according to Geological Survey of Canada reports. Survey personnel regularly visited the northern territory in the late nineteenth century, and it seemed likely that its rock formations were similar to those around Sudbury, Ontario, where mines had recently been opened to exploit copper and nickel lodes; it was therefore anticipated that the same ores would be found in Quebec. A mining project, however, would require trains to transport ore and settlers. It remained to be seen whether the availability and size of the deposits would justify such an investment.

Even if ore deposits were not found, the commissioner of Crown lands knew that the northern territory sat on a clay belt and might therefore be suitable for colonization. What is more, Quebec and Ontario were in negotiations with the federal government regarding the cession and division of this part of the territory, which had belonged to the Hudson's Bay Company.[40] With this in mind, in 1897 the commissioner of Crown lands sent Henry O'Sullivan, Quebec's inspector of surveying, "to make an exploration of the country between Lake St. John and James Bay" to assess the possibility of building a railway and determine a potential route.[41]

This exploration was part of a broader strategy by the provincial government, which was also making extensive use of publicity to attract capital and people to a space that it was gradually laying claim to and seeking to develop. Similar actions were taken by economic and political leaders in Quebec City, who were seeing their local economy deprived of a promising future due to the high level of activity at the port of Montreal, which had become the hub for the trade in grain from the West.[42] In the view of the businessmen in the Quebec City Chamber of Commerce and the members of the Société de géographie de Québec, who spoke to the same forums, the government had to support the plan of the Quebec and Lake St. John Railway to lay track across the territory newly acquired by the province and hope to find natural resources there. If the plan did not come to fruition – if only the "western tip" of this territory was served – "the vast resources of all the territory north of the height of land will remain dormant and of little or no practical utility to the Province."[43] Nor, one might add, would it bring profit to the port of Quebec City. It was with this in mind that O'Sullivan proposed that the National Transcontinental Railway run up the Nottaway River and along the shore of James Bay on its way westward.

Once the land formerly owned by the Hudson's Bay Company was ceded to Quebec in 1898 and, especially, once the mining rush took hold in northern Ontario – the discovery of silver deposits during construction of the Temiskaming and Northern Ontario Railway led to a major influx of prospectors to northeastern Ontario starting in 1902 – the Quebec Department of Colonization, Mines, and Fisheries sent new expeditions to assess the ore deposits in the environs of Lakes Mistassini and Chibougamau, as well as any other assets likely to encourage the establishment of a frontier settlement. The aim of the expeditions was to make visible the geological structure of "Nouveau-Québec" (the name first used for the northwestern part of the province) and its mineralogical potential from the Chibougamau district to Lake Témiscamingue, on the interprovincial border, as Geological Survey of Canada reports had repeatedly inferred that deposits being mined in northern Ontario extended into Quebec territory. In his first exploration report, O'Sullivan underlined the significance of cataloguing the resources "to enlist foreign capital" in proving that activity in the region constituted "a safe and solid investment" and was "connected with other interests of the highest national importance."[44] He therefore returned to explore the area a second time in 1898, following which he reiterated the need for the National Transcontinental Railway and the Grand Trunk Pacific Railway to adopt a more northerly route (see Figure 2.3).[45] In 1905, he received instructions from the minister to "verify the elevation of the watercourses and rivers of the Lake Mistassini basin and the Nottaway [River]," paying particular attention to "said country of Lake Chibougamou [sic], which is interesting from the point of view of mineral deposits that it contains and which our mining engineer, Mr. Obalski, has just explored."[46] Indeed, after receiving asbestos and copper samples from a fur trader, Peter McKenzie, who had gone to look for iron north of Lake Chibougamau, Obalski had convinced the minister of the need to verify "whether this region is truly fit to be developed."[47] Obalski was excited by the discovery of asbestos-bearing serpentines several kilometres long, and he hoped to start up a new mining centre in Quebec: "The results of my exploration have surpassed all my hopes and I am convinced that the province has a new and very important mining district, comparable to that in our Eastern Townships."[48] He hoped that the provincial government would quickly engage in construction of a railway to launch a mining industry as active as the one that had followed the discovery of silver in Cobalt, Ontario. Obalski led a second expedition around Lake Chibougamau in 1905 to update the maps and expand the province's mineralogical potential.[49] Finally, he

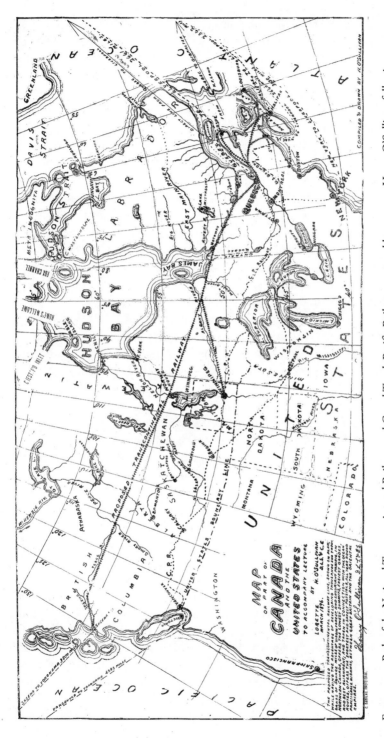

FIGURE 2.3 Path of the National Transcontinental Railway and Grand Trunk Pacific Railway proposed by surveyor Henry O'Sullivan following his second expedition to northwestern Quebec. | *Source:* Henry O'Sullivan, *Rapport préliminaire sur l'exploration de l'étendue de pays comprise entre le Lac St-Jean et la Baie de James* (Quebec City: Charles Pageau, 1899).

mandated Dulieux to investigate the appropriateness of laying track to link the Chibougamau district to the town of Roberval – a plan for which Dulieux concluded, upon completion of his expeditions, was difficult for him to supply "a positive opinion."[50]

Also in 1905, the Geological Survey of Canada once again sent Low to Lake Chibougamau, responding to a petition submitted to Prime Minister Laurier by residents of Quebec City that evoked "the newly discovered mining region."[51] Spurred by mining activity in northeastern Ontario, the Survey sent geologists Morley Evans Wilson and Alfred Ernest Barlow to the area east of Lake Témiscamingue to investigate whether the rock formations containing silver- and gold-bearing ores discovered in the Cobalt and Kirkland Lake districts extended into Quebec.[52] In 1907, Barlow, a former employee of the Geological Survey of Canada who had been exploring the Ontario side of Lake Témiscamingue since the mid-1890s, republished a report that he had written in 1899 on the geology and natural resources in the Nipissing district in Ontario and Pontiac County in Quebec.[53]

For both the Geological Survey of Canada and the Quebec Bureau of Mines, the extension into Quebec of the lodes that had led to "such marvellous discoveries" in Ontario was key to the fieldwork aimed at demonstrating the similarity of the formations on either side of Lake Témiscamingue.[54] Furthermore, the superintendent of mines (the official title for Obalski's position starting in 1906) was certain that the ore belt stretched up to Lakes Chibougamau and Mistassini, which, in his view, was its eastern edge.[55] The shaping of this mining space, "our great northern region, from Témiscamingue to Mistassini," mobilized the department's publicists, who eagerly anticipated a repetition of the Cobalt story in Quebec. Indeed, hundreds of mining companies had settled in northeastern Ontario in just a few years, and the prospecting rush had revealed other deposits – gold, this time – in Kirkland Lake in 1906. The discovery of a gold deposit at Porcupine Lake and Timmins, Ontario, in 1909 had a similar effect.[56] Quebec's Department of Agriculture then publicized the similarity of geological conditions on either side of the border and anticipated "important discoveries of minerals in the region to the east of Lake Abitibi traversed by the Transcontinental Railway."[57] The rapidity with which the Bureau of Mines latched onto this issue is clear when we compare two publications produced five years apart for the Paris (1900) and Liège (1905) world exhibitions. In the latter case, Obalski announced the discovery "in the district of Chibogomo [sic], in the northern part of the province, [of] a very extensive green serpentine band containing asbestos *analogous* to that in Thetford and Lac Noir ... If the prospecting expeditions organized this year result in other good discoveries, a railway will be built to develop this district, which contains other industrial ores."[58]

Hoping to attract prospectors to northwestern Quebec and create the equivalent of a "new Ontario" in the province, but with his hands on only some of the results of the Geological Survey of Canada explorations, Obalski sought to transform "Nouveau-Québec" into a burgeoning Canadian mining space. In his publications, he built analogies with the metal belt in contiguous New Ontario so that the reader would infer the presence of similar ores around Lakes Témiscamingue and Chibougamau.[59] For instance, in 1906 he shared a major discovery of ore in rock that "resembled the schistose diabase found in the Cobalt region."[60] With more difficulty, he drew a parallel between the asbestos veins found around Lake Chibougamau and those profitably exploited in the Thetford region.[61] Although Low was circumspect with regard to the value of the mineral wealth around the lake in his 1905 expedition report, Obalski reproduced in his own subsequent reports only the passages that confirmed his views and goals.[62] Indeed, Low was more cautious than were his Quebec counterparts for, in his view, without a railway, it was clear that ore extraction could not be profitable.[63] No matter; the surveying and mapping of the Lake Chibougamau region by the Quebec Bureau of Mines was accompanied by intensive prospecting activity between 1905 and 1910.[64] "What is believed to be reliable authority" claimed that 200 to 300 people per year, most from northeastern Ontario, were following the deposits across the interprovincial border to explore and prospect around Lake Chibougamau, convinced that the area was bound to contain more riches than the Cobalt region.[65]

For Obalski, whose Bureau of Mines was ensconced in the Department of Colonization, Mines, and Fisheries, the construction of a railway and the start-up of a mining industry would present the combined advantage of accelerating occupation of a territory linking two regions where colonization had recently advanced but that were separated by the terra incognita that the province had just acquired. These territories were not equally accessible in terms of the train tracks in place and to be lain. Furthermore, railway construction was a highly controversial undertaking, both politically and financially, as the province had fallen deep into debt during the track-laying frenzy of the previous decades. Finally, with equal mining potential, could the two regions be colonized at the same time? Would that not monopolize the limited resources that the government intended to make available to both the colonization movement and the mining industry? The latter was complaining about subsidies spent on colonization, which it regarded as being in competition for these resources. The backcountry of the Saint-Maurice Valley and Lake St. John, like the northern extremity of Pontiac County, offered few prospects for intensive farming activity, and what ores actually existed there was difficult to assess because little effort was being devoted to exploration and prospecting. It thus remained to be determined which of the two regions would eventually receive the attention not only of prospectors and mining syndicates but also, and most important, of the Bureau of Mines geologists and the railway companies that would bring the equipment and transport the ore.

In 1910, "on the recommendation of the newly-appointed superintendent of mines, Mr. Theo. Denis," the minister of colonization, mines, and fisheries, C.R. Devlin, mandated a commission "to investigate the natural resources (chiefly mineral), and to study the geology, of the Chibougamau Region," and to "answer not only the question as to whether the construction of a railway to Chibougamau is at present justified by reason of the mineral discoveries already made and partially developed."[66] The minister was reacting to reports that were somewhat contradictory and far from definitive with regard to the mineral assets in the area and to numerous requests from industrial and political quarters for construction of a railway that would facilitate the start of exploitation in the district. The mandate specified that, given the high level of speculation in recent years, the government hoped to reach a final conclusion with regard to the "potential of this region from the point of view of mining." The minister wanted to ascertain as precisely as possible the nature and wealth of the mines in the area. He specified that he "expected to find asbestos and gold" and that "if the expedition report guarantees the riches of the Chibougamau region, train tracks will be routed in that direction."[67] To this end, the Chibougamau Mining Commission was chaired by Barlow, who was now a consulting geologist in Montreal and special lecturer in economic geology at McGill University, "as it was desirable that the verdict of the Commission should be authoritative and so far as possible conclusive" (see Figure 2.4).[68] It also included a mining engineer, John Cole Gwillin, a professor at Queen's University, and Eugène Rodolphe Faribault, a cartographer with the Geological Survey of Canada.

In their final report, the commissioners concluded "that after carefully weighing the evidence which has accumulated as a result of their examination and study of the district, [we] cannot find that the mineral deposits so far discovered are of sufficient merit to justify the spending of public money in the building of a railway as proposed from Lake St. John to Lake Chibougamau."[69] Notably, the commissioners concluded that there was no foundation for likening the asbestos serpentines at Lake Chibougamau to those in the Thetford region.

The conclusions were as unequivocal for colonization of the region as for establishment of a mining industry. Furthermore, the construction of a railway was indefensible, economically speaking, in the opinion of the members of the Chibougamau Mining Commission.[70] In the Quebec Legislative

FIGURE 2.4 Members of the Chibougamau Mining Commission (Professor John Cole Gwillim and his assistant, A.M. Bateman) at work in the sampling mill in Pointe-aux-Bouleaux. | *Source: Rapport sur la géologie et les ressources minières de la région de Chibougamau (Québec) par la Commission minière de Chibougamau ...*, translated by J. Obalski (Quebec City: L.V. Filteau, 1912), 20.

Assembly, some saw the commission's findings as the government's justification for avoiding any commitment to another financially devastating railway and colonization venture. Soon after the report was tabled in the Legislative Assembly, the member from Charlevoix asked Minister Devlin if it were "true that the commission sent to Chibougamau by the current government to report on its value had been instructed before its departure to find a way to produce an unfavourable report in order to free the government from demands to build a railway between Roberval and that location."[71] Indeed, Obalski's retirement in May 1909 and his replacement by Théophile-Constant Denis, a former employee of the Geological Survey of Canada, in January 1910 provided an opportunity to dampen enthusiasm surrounding the mining potential of Lake Chibougamau and redirect prospectors' efforts to the northwestern part of the province. Not only was this region adjacent to the new mining region in Ontario, but it was in other ways better situated than Lake Chibougamau to profit from the coming National Transcontinental tracks. This railway project, which was to stretch from Quebec City to the Pacific, was already well underway, and the benefits seemed more certain for both colonization and industry if the line stayed north of Pontiac County.

What is more, the line would also serve the mining industry in Ontario – both its extraction sites and its main financial centre. Once a branch line was laid linking Rouyn to the Temiskaming and Northern Ontario in 1926, some in Quebec boasted of being able to reach Toronto, the true financial centre of the Canadian mining industry, by train in less than twenty-four hours.[72]

The Invention of a Mining Space

As all the work publicizing the deposits at Lake Chibougamau was washed away under the wave of the commission's conclusions, the Bureau of Mines directed its efforts toward the northwestern part of the province, between the height of the land and Hudson Bay. The laying of the National Transcontinental tracks accelerated colonization of Témiscamingue by opening up the region, and the centre of the Quebec mining industry shifted from south of the St. Lawrence River – where it had been concentrated during the second half of the nineteenth century – to Abitibi. This contributed to occupation of the territory and consolidation of the scientific activities of the Bureau of Mines.[73] Indeed, far from being the result of a chance discovery, the exploitation of copper and gold deposits was the object of intense scrutiny by the provincial agency starting in the early twentieth century. The goals were to stimulate colonization of the Abitibi plain and, especially, to diversify and intensify the province's mining industry, which was dominated by asbestos extraction and had been largely supplanted by mines in Ontario and British Columbia. After Denis was appointed superintendent of mines, the mission and means of the Bureau of Mines expanded as new ores were discovered, and revenues began to grow – "clearly a surplus over expenditures – $62,737 [in 1910] compared to $8,000 in 1908."[74] Under Denis, the bureau contracted with many more university geologists to explore the territory. Dulieux went to study iron ore in the Côte-Nord and Charlevoix regions in 1910, and in the Laurentians, the Saint-Maurice Valley, the Lac-Saint-Jean region, the Gaspé, and the Eastern Townships in 1911 and 1912.[75] A young graduate from the École polytechnique de Montréal, Adhémar Mailhiot, conducted geological reconnaissance in the York and Sainte-Anne river basins in the Gaspé in 1910.[76] After investigating granite deposits in the Eastern Townships for the Geological Survey of Canada, Mailhiot succeeded Dulieux at the École polytechnique, and in 1917 he returned to explorations for the Bureau of Mines, this time around zinc and lead deposits in Lemieux Township in the Gaspé. But it was mainly in northwestern Quebec, on land crossed by the National Transcontinental Railway tracks, that explorations financed by the Bureau of Mines multiplied, as did work done by the Geological Survey of Canada in the region during the same period. After Obalski's expedition to the northern part of Pontiac County in 1906, the explorations seem to have been punctuated by gold discoveries made by prospectors, successively, in 1906 in Boischatel Township, in 1910 in Duparquet Township, and in 1911 and 1912 in Dubuisson Township. Robert Harvie, a demonstrator at Harvard University and seasonal employee at the Geological Survey of Canada, travelled through Fabre Township, in the region east of Lake Témiscamingue, in 1910, before obtaining a permanent position at the Survey in 1912.[77] Finally, Joseph Austin Bancroft, a professor at McGill University, explored the region of Lakes Keekeek and Kewagama and that of the Harricana and Nottaway Rivers in Abitibi in 1911 and 1912, after gold was discovered on the banks of Lac de Montigny by prospectors from Ontario.[78]

The reports by these academics and Bureau of Mines employees who travelled through, probed, and mapped northwestern Quebec – including the region around Lake Chibougamau – between 1906 and 1912 did not simply supply a list of resources or a tool to help the government promote a mining region but also described the possibilities of access to a territory otherwise deemed impenetrable. Thanks to the tracings of access routes, prospectors working in neighbouring areas of Ontario had at hand a general guide for reaching distant stake areas. Before the National Transcontinental Railway was completed in 1912, the rivers described and used by surveyors and geologists formed the main travel network in the region for prospectors from the mining camps of Cobalt and Kirkland Lake. In fact, the first deposits to be claimed were in the region adjacent to these rivers.

Mineralogical determination was another activity that the Bureau of Mines subsidized with a view to improving and appropriating the territory. Indeed, when he was extolling to prospectors the services of an assay laboratory that would analyze and identify mineral samples "at a reduced price," Obalski specified that he wanted to "stimulate a greater number of analyses and encourage the discovery of ores" and expand the prospection zone.[79] It was even more important to push prospection because it was the only way to pinpoint the potential for exploitation of the subsoil and the longevity of future mines.

However, the Bureau of Mines did not step forward to perform this work itself, not only for lack of personnel and resources but because representatives of the mining industry considered it the prerogative of private enterprise. Through its mouthpiece, the General Mining Association of the Province of Quebec, the industry also demanded that a technician on the government payroll perform mineralogical determinations, which would stimulate prospection of a territory without hindering free enterprise.[80] In fact, the number of mineralogical determinations had grown significantly since sampling was handed over to the engineering consulting firm of Milton L. Hersey in Montreal: from 15.8 between 1897 and 1901, the annual average of number of analyses performed for the Bureau of Mines increased to 439.8 between 1902 and 1906, then to 1,511.33 between 1907 and 1911. In July 1911, the recently appointed director of the bureau, Denis, established the Laboratoire provincial d'analyse des minerais at the École polytechnique de Montréal, where new laboratories, directed by Dulieux, had been in operation since April 1908.[81]

The activities of the Bureau of Mines helped channel Ontario prospectors toward northwestern Quebec while engaging them in geological exploration and mineralogical determination aimed at increasing knowledge of the subsoil in that region. Despite this intensive mapping of the territory and the bureau's efforts to develop mine operations in the northwest, however, investments did not follow. To accelerate prospecting and mine operation among claims holders and dissuade them from simply speculating, the government tried, without success, to solve the problem through amendments to the General Mining Act in 1909 and 1910.[82]

Prospecting and geological exploration in the northern part of Pontiac County slowed temporarily during the First World War, but the first discoveries were confirmed when activity resumed after the end of hostilities. Notably, a prospector from northern Ontario, Edmund Horne, found copper and gold deposits near Lake Osisko in Rouyn Township in 1920, nine years after he had first prospected the site. A prospecting rush ensued, and geological explorers gradually returned to the region. The Bureau of Mines sent Mailhiot – now a professor at the École polytechnique de Montréal and director of the Laboratoire provincial d'analyse – to study the gold deposits at Lac De Montigny in 1921. During the same period, the extension into Quebec of the northeastern Ontario metallic belt was confirmed both by the Ontario Bureau of Mines and by the Geological Survey of Canada, which undertook a detailed mapping of northwestern Quebec in 1922.[83]

During the 1920s, a new prospecting rush along the routes established and the claims staked out before the First World War was accompanied by the formation of financial syndicates for the construction of mines in Abitibi.[84] The Noranda Mines operation went into production in 1927, on land staked out by Horne several years before, and more than thirty mines were opened in the region during the following decade. Copper extraction was central to mining activities in the Rouyn sector, but elsewhere – especially around Val-d'Or, the other mining hub in Abitibi – it was mainly gold ore that was extracted, as the price of gold rose in the context of an international economic crisis and monetary devaluation.[85]

In the northern part of Pontiac County, the government created the infrastructure needed for mining, as had been demanded by prospectors, explorers, and the political and economic elites of Quebec City and Lac-Saint-Jean region for the area around Lakes Chibougamau and Mistassini since the early twentieth century.[86] However, in the wake of the report by the Chibougamau Mining Commission, the Bureau of Mines stopped devoting its limited resources to this region in order to concentrate its explorations mainly in the northwestern part of the province. Then, when the mining rush took hold in Abitibi in the second half of the 1920s, the Bureau of Mines explored that region more systematically, just as the Department of Colonization, Mines and Fisheries was starting to pay greater attention to supervising mining operations. It established claims registration offices in the very heart of Abitibi, in Ville-Marie and Amos, in 1923 and 1924, respectively. It also financed two independent assay laboratories in Angliers and Amos in 1924. There was no longer any question of running things remotely from Montreal or Quebec City or of slowing mining colonization of this region: the Bureau of Mines set up premises on site to administer mining titles and analyze samples and to oversee the inventory of assets and the mapping of the area at a local scale. As it financed and supervised exploration in different parts of the region, the Bureau of Mines annually presented an inventory of gold and silver extraction in northwestern Quebec starting in 1922, then of ore deposits starting in 1925, and finally it started to map mine properties at regular intervals in 1933.

The creation of northwestern Quebec as a mining space (see Figure 2.5) made expansion of the Bureau of Mines possible, culminating in 1929 with the creation of the Division of Geology.[87] Although the superintendent of mines had reasoned that the bureau was saving money by hiring university professors during the summer – they received their main income from their institution the rest of the year – an amendment made to the General Mining Act in April 1929 invoked the need "to no longer depend entirely on temporary technicians for geological investigations, as was previously the case."[88] Whereas the 1880 statute had authorized the conduct of geological explorations, the 1929 amendment stated that "it would be permissible for the minister to create within the Bureau of Mines a geology and mineralogy agency whose functions would be to perform explorations and studies in the field to execute geological and mineralogical studies and publish maps and geological and mineralogical reports."[89] No longer was it simply a question of allowing explorations to take place; now, the Quebec state was to be provided with an appropriate institution that would develop "a formal exploration bureau."[90] Alphonse-Olivier Dufresne, who succeeded Denis as superintendent of mines when the latter resigned in 1927, piloted these

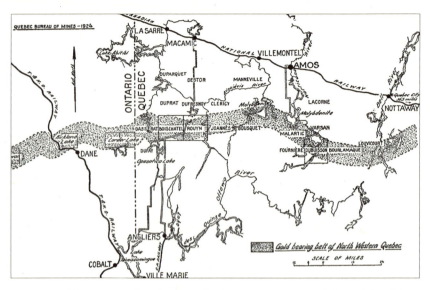

FIGURE 2.5 The creation of northwestern Quebec as a mining space. | *Source:* Théophile C. Denis, *Esquisse géologique et minéraux utiles de la province de Québec* (Quebec City: Ministère de la Colonisation, des Mines et des Pêcheries, bureau des mines de la province de Québec, n.d. [1924]), n.p.

changes, which the industry had been demanding since the early 1920s.[91] A graduate of the École polytechnique and McGill University, Dufresne had participated in the work of the Chibougamau Mining Commission in 1910 and in Bancroft's explorations in 1911 and 1912, before joining the bureau as assistant superintendent in 1914. Finally, one year after the Division of Geology was created, the growing autonomy of the mining sector within the public administration was manifested by profound institutional change. On April 4, 1930, the Bureau of Mines was placed under the aegis of a cabinet minister devoted solely to the sector, Joseph-Édouard Perrault, who had been minister of colonization, mines, and fisheries since 1919, under Premiers Gouin and Taschereau.[92] The industry, which had been demanding this appointment for ten years, quickly and repeatedly expressed its satisfaction.[93]

Leading the Division of Geology was a consulting geologist, John A. Dresser, who, for a dozen years, had conducted explorations in the Eastern Townships for the Geological Survey of Canada. The division undertook to intensify geological exploration, mineralogical determinations, and mapping. Dresser's first action was to hire three geologists, Bertrand-T. Denis, Islwyn Winwaloc Jones, and Leslie V. Bell, all of whom were graduates of an engineering or mining school and had graduate degrees. Jones and Bell had work experience in a state organization – the Geological Survey of

Canada and the Ontario Bureau of Mines, respectively – and Denis had worked with Bancroft for the Anglo-American Corporation of Northern Rhodesia.[94] Along with centralization of the other technoscientific activities of the Bureau of Mines (see below), the explorations conducted by the division's personnel provided a basis for the Quebec state's control over the province's geology. As political scientist Richard Brunelle notes, "the permanent geologists themselves led most of these expeditions and coordinated all of the geological work, such that it was possible, for example, to determine which regions were a priority to cover."[95] Between 1929 and 1939, the division organized 104 geological expeditions, whereas there had been only 12 between 1906 and 1929 (see Figure 2.6).[96] The purpose was not to assemble a systematic geological record of the province as a whole, but to interpret the geology of localities through "analysis of the nature, distribution, and structural relations of mineral formations" and, in particular, to "indicate formations appropriate for mineralization."[97] This intensification of exploration went hand in hand with expansion of mining production in the 1930s, despite the worsening economic crisis; extraction of metallic ores in Abitibi drove the province's mining industry and encouraged the Quebec government to open new offices to serve mine operators.[98]

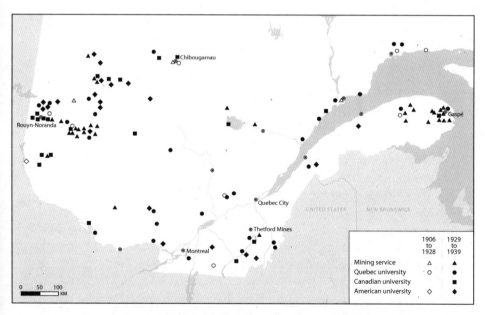

FIGURE 2.6   Sites of explorations by the Quebec Bureau of Mines (1906–39) and institutional affiliations of geologists. | *Sources: Reports on Mining Operations in the Province of Quebec, 1906–29; Annual Reports of the Bureau of Mines, 1929–36; Geological Reports, 1937–40.*

The Bureau of Mines was still counting on professors at McGill University, Université Laval, and the École polytechnique de Montréal to survey the territory, but it tightened its hold over this activity as the number of expeditions led by its own employees rose.[99] Students from Université Laval, McGill University, and the École polytechnique accompanied the bureau's geologists on their explorations, allowing for future personnel to be trained to suit the needs of the state institution.[100] Some students from Ontario and American universities even chose to complete their master's theses or doctoral dissertations at the bureau, under the supervision of Dresser or of T.-C. Denis, who remained actively involved.[101] This resulted in a strengthening of the geologists' position within the department and in the state's taking charge of the province's mining territory and its exploitation.

Cartographic production by the Bureau of Mines illustrates this repositioning of technoscientific personnel, as can be seen from a comparison of three mineral maps, published in 1914, 1924, and 1930. The two earlier maps were made similarly, and they are distinguished only by updating to include new ore deposits. Reproduced from a book published by the Geological Survey of Canada, they are found in publications by a single author, T.-C. Denis – the first in an article describing the province's geology and mining industry in the first edition of the *Annuaire statistique de la province de Québec* (see Figure 2.7) and the second in *Esquisse géologique et minéraux utiles de la province de Québec*, a monograph produced for a general readership at the British Empire Exhibition held in London in 1924.[102] Although both publications contained a geological map, they presented, in addition to the ore sites, much topographic and toponymic information and careful tracings of the many railways' train tracks. The ore deposits were indicated in red, but they were drowned in a sea of details that also included names of watercourses and towns, along with political borders. As a result, the portrayal of potential mining sites in Quebec, widely scattered through the province's vast territory, seemed to suffer from a desire for scientificity that prompted the Bureau of Mines to mimic the Survey's publications.

With the map published in 1930 (see Figure 2.8), the Bureau of Mines sought to make up for this lack of legibility. The scale was the same, but the watercourses had basically disappeared (as had the Magdalen Islands, for which was substituted a deposit of manganese in the middle of the ocean!), except for those situated close to ore deposits. The only cities to appear were Ottawa, Montreal, and Quebec City, but the author of the map took care to write in capital letters the names of four mining regions, two of which – Chibougamau and Gaspé – were still in a hypothetical state of industrial development, compared with the other two – Rouyn and Thetford – where the province's mining production was concentrated. Finally, the main railway

The Invention of a Mining Space

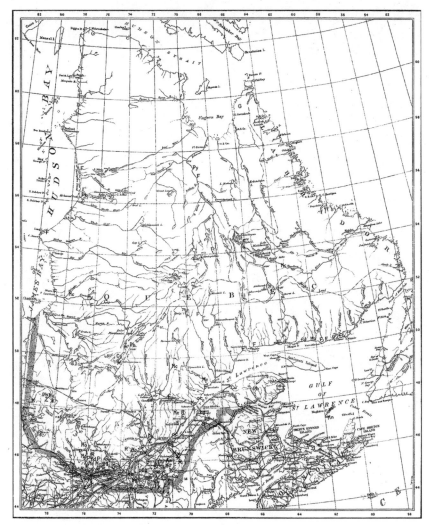

FIGURE 2.7 Mineral map of Quebec, 1914. | *Source:* T.-C. Denis, "B-Esquisse géologique," in "Description de la province," *Annuaire statistique de la province de Québec* (1914), 16.

lines were retained, but, where the reader might doubt the accessibility of these tracks, the author of the map cross-hatched the lines to make it clear that they were there to reach deposits or transport the ore extracted.

The 1930 map had many goals aside from showing the rise of gold and copper mining in the northwestern part of the province since the publication of the 1924 edition of the map. Of course, numerous other deposits had been discovered through explorations that the Bureau of Mines had conducted in the intervening years, but the bureau wanted first to establish clearly that the

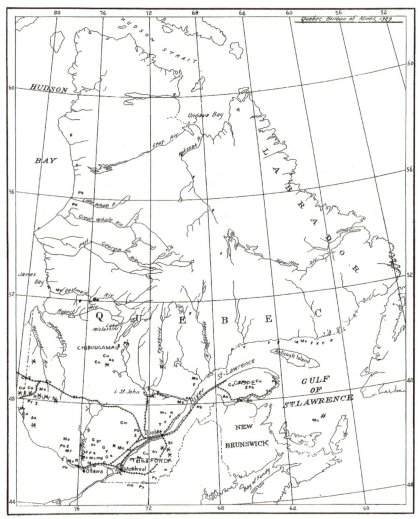

FIGURE 2.8 Mineral map of Quebec, 1930. | *Source:* Quebec (Province), Bureau of Mines, *Mineral Map [of the Province of Quebec]* (Quebec City: Bureau des mines, [1929]).

province belonged to the mining industry through the production of these mining spaces and to promote their mineralogical potential. Second, it was no longer inspired – directly, at least – by cartographic production by the Geological Survey of Canada, as it was no longer reproducing the Survey's maps by rote. Finally, there was no longer a question of confusing economic information with purely scientific data in a single map. In other words, the map signalled the independence of mining concerns from the strictly geological aims of the technoscientific personnel of the Bureau of Mines. In fact, unlike the preceding publications, the 1930 publication did not include a geological map, and, over the following decades, the province's geology as a whole ceased to be mapped, except for a few sketches made freehand. Only certain regions had their geology displayed, and at a finer scale, to allow readers to better grasp the rock formations and mineralization zones that were particularly interesting for mining operations. These representations were the result of numerous expeditions launched annually by the new Division of Geology, to which the bureau added a Division of Cartography in 1934.

By holding the fort on exploration, the Bureau of Mines consolidated its appropriation of the territory and its relationship with the mining industry. To this end, the Division of Geology concentrated most of its personnel in the Abitibi region, sending reconnaissance missions to the edges of this mining ecumene in Pontiac and Témiscamingue counties and around Chibougamau, which "reappeared" in Quebec's mining space in the early 1930s as prospecting and claims registrations resumed. No regions were neglected, however, whether they were close to urban centres, notably for quarries for granite and other construction materials, or in the Lac-Saint-Jean region, the Laurentians, the Gaspé, and the Eastern Townships – the last being the only area other than Abitibi where mining was profitable.

Finally, scientists' increasing influence on the organization of their work was manifested not only in the definition of exploration activities but also in the push to centralize analysis and mineralogical determination activities. Despite a rise in mining production in northwestern Quebec, the Bureau of Mines stopped using the private laboratories in Rouyn (previously in Angliers) and Amos that it had been financing since 1924, repatriating analysis activities to its Laboratoire des mines in Quebec City, which began operating in September 1933.[103] In December 1936, the bureau brought the various analysis and determination activities into a new complex in Quebec City, as the laboratory was involved more and more in serving the growing needs of geologists, of whom the bureau was sending ever greater numbers into the field.

BY DEFINING THE TERMS for appropriation of the subsoil and exploitation of mineral resources, the General Mining Act of Quebec made the growth of provincial geology possible. Analysis of the scientific activities coordinated by Obalski reveals, however, that information about the territory's mineral resources usually came through outside sources, especially the Geological Survey of Canada. This information proved insufficient to stimulate the growth of a mining industry, and in the wake of the first discoveries of deposits in the northwestern part of the province, the Quebec Bureau of Mines undertook to build its own expertise, taking over from the Geological Survey of Canada as the principal repository of knowledge concerning the province's subsoil. These developments had an impact on the distribution of personnel for more than thirty years, resulting in a concentration of geological exploration in certain places to the detriment of others, due to limited resources or to the desire to intensify reconnaissance of a specific space and publicize its mining potential once mapped. The changes also resulted in consolidation of the technoscientific staff at the Bureau of Mines and an increased number of explorations for the purpose, first and foremost, of revealing the geology of the northwestern part of the province, determining the terms of its occupation, and orienting mining operation in that region.

Although it was not the only factor, the Bureau of Mines, through its scientific activity, contributed to how the Quebec state apprehended occupation of its territory and exerted its government of mineral resources. Of the two hypotheses advanced to explain the development of mining in Abitibi – the arrival of the railway and the wealth created by copper and gold extraction – it must be noted that the railway preceded the gold rush by fifteen years and the wealth came after. Another structuring factor was the geological exploration and mineralogical construction of the region, the pace and forms of expansion of which have been discussed here. Finally, the autonomy of researchers in the definition of their research activities must be emphasized. Despite the concentration of exploration and mapping activities in the northwestern part of the province, the Bureau of Mines sought to deepen its knowledge of the province's geology by exploring the Gaspé and the Côte-Nord region despite the absence of a mining industry in these territories, and the Eastern Townships, where a declining copper industry was subsisting. As for the prosperous asbestos-extraction industry, it received no specific attention from provincial geologists after the early twentieth century, except for an updating of the surveys produced at the turn of the twentieth century and the creation of a laboratory for standardizing the quality of the fibre.

3
Soil Classification and Separation of Forest and Colonization Areas: Scientific Forestry and Reforestation

IN THE FIELD BOOKS compiled during their expeditions, the Department of Crown Lands surveyors listed mineral deposits, animal species, water power, tree species, and the agricultural potential of the soil. Of course, the department wanted to know about the topography and resources of the land obtained under the British North America Act, but it was actually essential to have this information to formulate coherent official descriptions of the timber limits that it was leasing to lumber companies: in return for the payment of an annual rent and stumpage fees, the companies obtained exclusive permission to cut the timber in their limits. For the commissioner of Crown lands, it was important, notably, to know the precise borders of these limits and to have information on the rivers flowing through them, to which the holders of timber limits had access to float their wood.[1]

A new issue arose in the late nineteenth century, when the department wanted to possess the knowledge necessary to separate land depending on its potential for timber cutting or agriculture and to define parcels of land to which settlers could move on the frontier – precisely where the lumbermen were moving their logging camps. Even as surveyors were dividing the territories newly acquired or in the process of settlement into lots intended to "open [them] to colonization and industry" in the late nineteenth century, the government was trying to quell a conflict between the colonization movement and the forestry industry by different measures – first legal, then technoscientific.[2] In fact, settlers and lumbermen were in competition to get their hands on a resource that was increasingly coveted as it seemed to become rarer. In the previous few years in the United States, the booming construction industry and expanding railways had raised the prospect of rapid depletion of the forest, causing anxiety that wood-based industries would have trouble finding the materials that were the basis of their revenues.[3] It was in this context that conservationism – a movement demanding scientific and state-controlled exploitation of natural resources – began to take shape in North America.[4]

The conservationist movement had its proponents in Canada, where politicians and wood dealers were closely following the American debate around the "timber question," especially after exports of squared timber tumbled due to the depletion of large pines in the Saint-Maurice Valley and, to a lesser extent, the Ottawa Valley.[5] In Quebec, contributions to the public purse from forest exploitation were of great concern to the government, as between 19 and 33 percent of its current revenue came from the industry in the last quarter of the nineteenth century. It therefore attacked these issues by modifying its forest regime – the terms under which it sold, and later leased, the public forest.[6] When the commissioner of Crown lands suggested that the forest be put into reserves as a way of protecting the wood resource – whether by preventing fires or by impeding overcutting – this was clearly seen as one way to contain and oversee the colonization movement. It so happens that it was a highly profitable approach for the government coffers, as the value of the timber limits located in the reserves was thus increased.[7] Finally, this form of administration of Crown lands and forests encouraged development of the paper industry, which succeeded the lumber industry as the main economic activity in the sector.

In this chapter, I examine certain *dispositifs* instituted by the Quebec state to keep settlers away from the forest and enable Crown lands to continue to generate an economic rent for the public treasury, all while stimulating industrial development in the province. Although initially reserves and national parks were central to governmental strategies, the creation of a corps of civil servants trained in a provincial school of forestry later offered the public administration a new lever for shaping the forest landscape. The use of scientific forestry – and, more precisely, of reforestation techniques and specific tree species – seemed to be pivotal to the government of the forest and its resources. This was the case notably when the provincial Forest Service turned to reforestation in order to return part of the territory abandoned to the colonization movement to productive use and to create a separation between forestlands and agricultural lands.

Forest Reserves and Colonization Lands

The surveyors' work involved the delimitation of timber limits; these spaces were the subject of lively debate starting in the late nineteenth century. The argument was mainly between lumbermen, on the one hand, and settlers and advocates of colonization, on the other. For the latter, access to the wood in these limits represented secondary income that was often crucial, whereas the former were jealously guarding their prerogative to exploit the wood on their leased parcels. Unhappy about being confined to the forest regime dictated by the holders of timber limits, the settlers made headlines by felling trees illegally. They also drew the fire of the conservationists, who, as "protectors of nature," considered them to be destructive agents who neither knew nor appreciated the value – economic or aesthetic – of the forest. It should be noted that it was easier to denounce the negligence of the settlers who cleared the forest and caused brush fires than to understand the rebellious acts of individuals excluded from the territories leased for timber cutting.

Promoters of the forest industry (holders of timber limits, merchants, and politicians) and conservationists, many of whom came from the same quarters, had a choice platform at the second congress of the American Forestry Association, held in Montreal in August 1882. Organized by Montreal timber merchants James and William Little, the congress was held on the theme of exploitation and conservation of forest resources.[8] Federal and provincial political authorities heard the message loud and clear and quickly adopted measures designed to respond to the needs expressed with regard to forest conservation. Classification of forestlands into lumbering and colonization areas and fire prevention were issues that particularly drew the attention of the Quebec commissioner of Crown lands, William W. Lynch. In addition to hiring forest rangers to patrol timber limits and apply the land and wood regulations, Lynch tried to kill two birds with one stone by creating a vast forest reserve near the tributaries of the Ottawa River in 1883 and other reserves in parts of Beauce, Compton, Wolfe, Arthabaska, Mégantic, and Dorchester counties the following year.[9] The goal was to guarantee a constant supply to timber merchants and regular revenue to the provincial government. Lynch also prohibited settlers from access to these forested areas, as they had been blamed either for clearing land and then abandoning it or for causing fires through uncontrolled burns. Before being put in reserve, however, the forest space had to be inspected to determine the availability of the wood resource to be exploited and the parcels inappropriate for agriculture. During the ten years that the inspections were ongoing, no sale of lots to settlers could take place. Reservation of a territory for the exclusive use of a single group raised the ire of colonization advocates. This dissatisfaction, among other things, swept the Parti national into government in Quebec City in the next election, and the new premier, Honoré Mercier, abolished the policy of forest reserves in 1888 and forbade holders of timber limits to cut wood on lots allocated to settlers but situated within their limits.[10]

The reserves reappeared in the province's forest landscape a few years later, in 1891, when the Parti conservateur returned to power. Inspired by the experience in Ontario, where Algonquin Park was founded in 1893 in order to conserve old-growth forest, maintain water levels, and provide a recreational area, Lynch established Parc de la Montagne-Tremblante in 1894 and Laurentide National Park the following year,[11] with the objectives of protecting forest and wildlife resources and regulating hydrographic basins.[12] W.C.J. Hall, the superintendent of parks – who was also responsible for fire wardens – noted that in the Laurentide National Park, where more than a dozen major rivers had their source, there was no question of banning timber cutting. On the contrary, he maintained: "2,300 square miles are under licence and some of these limits are being operated at the present time, e.g., in the valley of the Batiscan."[13] In Hall's view, the use of forestry methods – including the removal of mature trees and the felling of softwood species only – could help to improve the forests. In addition, he noted that, thanks to the forest protection made possible by Laurentide National Park, wood floating could continue in times of drought, "contrary to what is observed south of the St. Lawrence."[14] He was referring to the fact that south of the St. Lawrence the land had been cleared and settled, whereas the territory put into reserve on the north shore could not be sold in lots to settlers and would therefore be protected against "reckless destruction." Thus, the parks had to be understood above all as spaces from which settlers were excluded in order to protect forest, hydraulic, and wildlife resources (see Chapter 4) and holders of timber limits could continue their tree-felling activities.

The advent of a pulp and paper industry heightened the dispute between settlers and holders of timber limits over access to the public forest and appropriation of wood on Crown lands. Not only did strong demand from the paper industry raise the value of the timber resource, but this new industry – even more voracious than the sawmill industry – increased the need to maintain, if not improve, the productivity of the forest.[15] The conflict became even more pronounced when the Quebec government sought to prevent exportation of pulpwood in order to stimulate pulp and paper production in the province: an embargo was instituted in 1910, ten years after a similar measure was adopted in Ontario.[16] However, the government had no control over private land, where "bogus settlers" – holders of location tickets with no intention of living on their lots – cut timber that companies bought and exported, to the detriment of timber limit holders who would have liked access to this wood to supply their market.[17]

In an attempt to mitigate these antagonisms, the provincial government created the Colonization Commission in 1902.[18] The commission's mandate was to "investigate the number and causes of conflicts between settlers and timber-limit holders, and to advise on how to prevent and quell such conflicts."[19] The commissioners reiterated the need to divide up the land according to its colonization or lumbering "vocation." Their report supported the view of those who for several decades had been demanding a definitive division of Crown lands through systematic soil classification. This issue was now even more pressing because the terms of the debate between the colonization movement and lumbering interests had evolved due to the rapid installation of the pulp and paper industry in Quebec, which was exerting strong pressure on forest resources to supply its mills. Because the industry cut intensively and was less discriminating about the tree species that it consumed, entire timber stands disappeared and the consequent denuded landscape was more pronounced and visible.[20] Similarly, the multiple roles of the forest became clear when excessive deforestation around drainage basins gave rise to irregular river flow, a problem for riverine residents affected by floods and for businesses that depended on the rivers for the flotation of wood or for production of hydroelectric power.[21]

The solution advocated to resolve the conflict remained division of the territory into lots for colonization and for the forest industry, and members of the Colonization Commission demanded that science be used to supply the data necessary for soil classification. Classification proved to be a complex undertaking, however, as it necessitated a detailed survey of Crown lands and scientific skills that the provincial administration did not have available in the early twentieth century. It was only after the Department of Colonization became responsible for monitoring timber cutting on colonization lands in 1921, followed by the creation of a bureau for the classification and use of forest soils in 1924, that the division between forestlands and lands suitable for agriculture was definitively instituted.[22] In the meantime, the Department of Lands and Forests continued to tacitly divide the territory by putting Crown lands into reserves.

In continuity with the national parks policy, the Department of Lands and Forests created forest reserves soon after the Colonization Commission's report was submitted in 1904.[23] Lomer Gouin, who had just replaced Simon-Napoléon Parent as premier, had previously argued in favour of the reservation of public lands as head of the Department of Colonization; he considered reserves "the only means to end the conflicts that arise between settlers and licence holders."[24] The "Gaspé forest, game, and fish reserve" was created by order-in-council on April 28, 1905. Then, to "increase the forestry value of the region and preserve river beds and sources," the Saguenay and Labrador reserve and the Rimouski reserve were created in 1906, and the Ottawa, Saint-Maurice, River Ouelle, and Chaudière forest reserves in 1907.[25] When the superintendent of parks suggested that a reserve be created in the Eastern

Townships, where private forests predominated and the area of the public domain was shrinking, the department evaluated the possibility of creating a reserve on the border with the states of New Hampshire and Maine, but ended up dropping the plan. This setback did not keep Premier Gouin from boasting to an audience composed of members of the Canadian Forestry Association that "Quebec has set apart more territory for forest reserves than the entire American Union has similarly reserved."[26]

The creation and mapping of forest reserves (see Figure 3.1) signified and publicized the end of the myth of the inexhaustible forest, making it even more urgent to contain the colonization movement. Not only did the map show the northern edge of the forest zone beyond which it was impossible to find the resources necessary to supply the pulp and paper industry, but it revealed that what remained of the exploitable forest was now in reserves – that is, in areas in which exploitation was controlled. The limits of the forest capacity in the province were therefore highlighted.[27]

In these reserves, the Department of Lands and Forests hoped that the forest would be exploited according to a pre-established working plan aiming for resource preservation. In practice, the forestry industry used both reserves and parks; timber limit holders continued to enjoy the privileges conferred upon them by their leases without fear of encroachment by settlers, as no lots were for sale in these reserves. Notably, the superintendent of reserves attributed the absence of forest fires to the withdrawal of lots for sale and the resulting absence of settlers.[28] The regulations in force allowed timber cutting while ensuring "the maintenance of natural irrigation as it exists at present and which is necessary for the most successful prosecution of the agricultural industry and for the protection and perpetuation of the fish and game in said region," even though the superintendent of parks recognized that there was "at the present time it is quite true ... no danger to be apprehended of lack of timber or water."[29] This reasoning mimicked the American conservationist rhetoric and discourses on soil desiccation. Preserving water volumes and guaranteeing regular flow for industrial hydroelectric power quickly became the main argument used by proponents of forest reserves, who vaunted their role in the preservation of riverbeds and sources at a time of strong expansion of the hydroelectric industry in Quebec. Furthermore, the creation of forest reserves was repeatedly presented as a measure aiming not to prohibit timber cutting but to encourage the rapid removal of mature trees to clear the forest floor so that growth of young trees would be accelerated and water volume in the rivers maintained.[30]

Even as he was promoting the forest reserve policy of the Department of Lands and Forests, the superintendent of parks took care to make clear that millions of inventoried and surveyed acres remained available to settlers

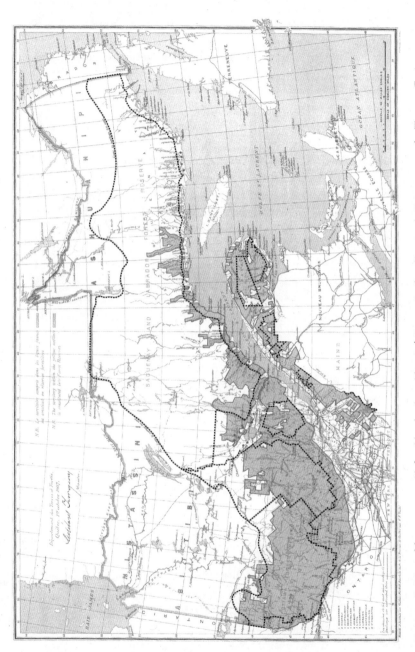

FIGURE 3.1 Map of forest reserves (defined by the dotted lines). | *Source*: Québec (Province), Département des Terres et Forêts, *Carte des territoires sous permis de coupe de bois* (Quebec City/Montreal: Département des Terres et Forêts/F.H. Denison Lith., 1907).

outside of the reserved zones.[31] This seemed insufficient, however: several years after establishing the network of reserves that covered the immense forested territory of the province, the department defined other territories designated for the exclusive use of settlers, who were otherwise forbidden access to the newly created reserves, timber limits, and parks. This was the objective of the township forest reserves created in 1911, when the department was planning to set aside barren and vacant land in the colonization regions. Under the supervision of forest rangers, these reserves were to provide village inhabitants with wood. The department also hoped to increase settlers' determination and desire to protect the forest by teaching them directly about forestry and encouraging a sense of common ownership of the forest.[32]

With the township reserves, the Department of Lands and Forests provided a new basis for its discourse with regard to settlers, at the same time as it was continuing to separate forest and agricultural areas. The plans were found in the Colonization Commission's report, which presented the paper industry as "the natural auxiliary" of colonization, as "it provides settlers with a means of making money with small wood from clearing that otherwise they would be forced to burn on the spot to clean their land."[33] In this way, settlers were encouraged to relate to the forest in a new way, as previously the wood cleared had been sold solely as firewood or lumber. In the view of the commissioners, "it is in the interest of colonization to stimulate the development of this industry as much as possible."[34]

Like the parks, the township reserves increased the area of protected forest space. The Department of Lands and Forests closely managed the township reserves – unlike their industrial counterparts, the timber limits – with regard to both inventories and supervision of timber cutting by department agents. Furthermore, the department's Forest Service, set up two years before the first township reserves, changed the function of these reserves in the colonization landscape by supplying the personnel and materials for silvicultural work in them. Reforestation, which began in the township reserves in the 1920s and accelerated during the 1930s, led to production of a province-wide pulpwood forest. Not only were the trees grown on land allocated to colonization, but the economic function of the forest territory reserved for settlers was diverted for the use of paper mills. In addition, the department banned "the felling of certain valuable trees, such as pine and spruce," in the township reserves when it was "not satisfied with their proportion or distribution in the forest."[35] Not only did reforestation play a key role in the process of reclaiming forestland from the colonization movement, but it constituted the keystone for introducing scientific forestry into Quebec and for training qualified personnel within the Department of Lands and Forests.

Scientific Training, State Forestry, and Reforestation

As the debate on forest depletion was taking place, voices were raised to draw attention to the need for scientific forestry, whether for soil classification and separation of agricultural and forest areas or for assessment of the availability of the wood resource on public lands that the province leased to lumbermen. These appeals ran headlong into the provincial adminstration's inability to act. The Department of Lands and Forests both was undermined by corruption and suffered from a dearth of skilled personnel, as the scientific apparatus was kept at a minimal level. How could the department take action if most of the technicians in its employ were friends of the regime and learning on the job? The problem was an absence of qualified personnel and the lack of means to correct this deficiency of the public administration.[36] What is more, unlike geology and agriculture, forestry had no equivalent to the Geological Survey of Canada or the Experimental Farms Branch on which the provincial Department of Lands and Forests could rely to acquire technical skills. The federal administration had had a forestry service since 1899, but it concerned mainly Crown lands in Manitoba and the Northwest Territories, where, as the entity responsible for administration of reserves, it encouraged the plantation of windbreaks and maintenance of farm woodlots.[37] The situation in Quebec was not exceptional, as Canada had no forestry school. In the United States, two schools had been established not long before at Yale University and Cornell University.[38]

In the first decade of the twentieth century, however, several provinces instituted forestry training. With the support of their respective provincial governments, the University of Toronto and the University of New Brunswick each created a forestry school in 1907 and a forestry department in 1908.[39] A few years before, the mining school at Queen's University in Kingston had considered offering similar training in Ontario, and in 1903 the university invited the former chief forester of the United States, Bernhard Fernow, to give a series of lectures with the intention of starting its own forestry school.[40] Trained in the great forestry schools of Prussia and founder of the New York State Forestry College at Cornell University, Fernow was an unconditional advocate of scientific forestry training.

In Quebec, Gustave-Clodimir Piché, a civil engineer working at the Belgo Pulp and Paper Company in Shawinigan, tried to convince Premier Gouin to create a provincial forestry school. Gouin, who had had Fernow's lectures – originally published in English – translated into French, turned to J.-C.-K. Laflamme, professor at and former rector of Université Laval.[41] An active member of the Canadian Forestry Association since its foundation in 1905,

Laflamme was concerned about exploitation of natural resources and saw the introduction of scientific forestry in Quebec as a way to stop the waste engendered by the timber companies' felling methods.[42] Similarly, science seemed the means most likely to resolve the most important "timber question" – "that regarding the relations that necessarily exist between colonization and the conservation of our forests."[43] In response to Gouin's request, Laflamme inquired about the training programs at the École des eaux et forêts de Nancy and the Yale University Forest School from their respective directors. He was hoping to send students to these institutions so that they would found Quebec's first forestry school upon their return.[44] In October 1905, at Laflamme's suggestion, the government sent Piché and Avila Bédard, a graduate of the Séminaire de Québec, to Yale University; in exchange, Piché and Bédard agreed to work for the Department of Lands and Forests when they returned to Quebec.

Reforestation of timber limits was a question to which Piché and Bédard paid sustained attention during their stay at Yale University, even though the few mentions of the subject in the scientific journals in the United States at the time would indicate that it was not a central issue in American forestry circles.[45] Piché and Bédard's research was stimulated by the constant demands by the Department of Lands and Forests, itself being pushed by paper companies with facilities in Quebec. Notably, the Belgo Pulp and Paper Company, where Piché had worked before going to Yale, was interested in methods for starting plantations in its limits.[46] At the request of the minister, Piché and Bédard wrote a report in which they recommended the use of red spruce seeds and detailed the methods for gathering them in the limits and sowing them.[47] Piché later sent instructions regarding future forest plantations directly to the chief of Belgo's forest operations; he advised "leaving the spruce trees in the nursery for three years before planting them permanently."[48] Fearing that Belgo would sabotage this plan by transplanting the seedlings too early, however, Piché recommended that the government "establish one or several tree nurseries in our province, designed to meet public demand."[49] He also believed that the availability of seedlings would encourage other companies to follow Belgo's example. As he anticipated "using [a surplus of trees] to reforest barren land or timber limits destroyed by forest fires, or to stabilize drifting sand like that found along the Canadian Pacific tracks, in the environs of Lanoraie," Piché suggested that the nursery be created as soon as possible "in order to be ready to meet public demand and start reforestation activities ourselves in the near future."[50]

Piché finalized the tree nursery project when he returned to Quebec after graduating from Yale. One of his first initiatives as a forest inspector for the Department of Lands and Forests was to travel along the north shore of the

St. Lawrence River, from Lavaltrie to Berthierville, from Lanoraie to Joliette, to find a site for the planned nursery. He submitted a report to the minister concerning the establishment of a nursery for forest trees, accompanied by a letter from Laflamme, who "strongly approve[d] of this suggestion."[51] Aside from cultivation and acclimation of tree species for individuals who wished to set up plantations, Piché proposed to reforest 25,000 "bare and sterile" acres with white pines to make woodlots. He also mentioned that he had held talks with the settlers and had "observed a keen interest in what, for them, was a new idea of leaving portions of their lots uncultivated for the production of wood."[52] In the view of Piché – and of Laflamme, who had pronounced himself in favour of the idea in 1906 – experimenting with large-scale reforestation was the responsibility of the state, rather than of individuals. Moreover, such activities would extend the policy of forest reserves launched in 1905 to counter the effects of deforestation.[53]

In autumn 1907, the department acquired the land that had been identified in Berthierville, and soil preparation, seeding, and plantation of seedlings began under Piché's direction in spring 1908. This nursery, however, was intended to do more than simply produce trees. As Laflamme had wished, Bédard and Piché formed "the nucleus of an education in forestry."[54] In 1909, Piché recruited eight students to "help with spring work." Some came from the École polytechnique de Montréal and had worked for him on soil classification and surveillance of forestry activities the previous year, and the others were graduates of the natural sciences program at the Séminaire de Québec. There were many tasks to perform at the nursery: selection of the seeds purchased, germination tests, digging, surveying, clearing land, wood scaling, breakdown, wood estimation, spading of borders, and transplantation of the previous year's seedlings. Piché explained to the minister that he gave them "very little theory, just what they need to explain the work that we are doing."[55] Nevertheless, this initiation was part of Piché's plan to found a forestry school at Université Laval, to which Laflamme had returned as rector in 1908.[56] Piché saw work in the nursery as an opportunity to recruit and select the first students for the school. As he explained to Laflamme, "if there are hopeless individuals in the group, we will await them next August, when the entrance exams take place."[57] When the École de foresterie finally opened at Université Laval in autumn 1910, Piché became its director, and he exempted the students who had worked at the nursery from one year of their study program.[58]

The creation of the forestry school went hand in hand with a reorganization of the Department of Lands and Forests. The department's personnel, concentrated in the Woods and Forests Branch, had become known for its lax oversight of lumbermen. Despite his devotion to the cause of the forest – he

FIGURE 3.2 The first students at the Université Laval forestry school received their training at the Berthierville nursery, before the school officially opened in Quebec City. | Berthierville nursery: students depart to conduct inventory and study the nearby bogs, 1909. Source: *Rapport annuel du ministère des Terres et Forêts* (*RAMTF*), 1909–10, 54.

had, among other things, overseen the setting up of forest reserves and national parks – the branch chief, Jean-Chrysotome Langelier, could not crack down on forest wardens because they were "protected and supported by the elected members in their ridings."[59] In any event, these wardens, politically appointed seasonal employees who had no skills appropriate to the forestry sector, simply neglected to monitor timber-cutting activities or collect stumpage fees.

As he was implementing his plan to train forest engineers, Piché wanted to reform the Woods and Forests Branch, to which he was attached as an inspector. He gathered several students around himself and Bédard in a forestry service, "the name given to the corps of agents and wardens working under the direction of the Forest Engineer and his assistant on exploration, classification, protection of Crown lands, control of lumbering operations, and preparation of accounts for timber cut illegally."[60] What distinguished these agents from the personnel in the Woods and Forests Branch was their training. The Forest Service that Piché led in 1909 comprised a group of technicians experienced with methods of scientific forestry learned at the nursery (see Figure 3.2). In 1910, after Langelier's death, Piché began to extend the service's operations throughout the province, and at the same time he centralized management of activities in the Quebec City office. His scientifically trained young officers gradually displaced the rangers of the Woods and Forests Branch, who were downgraded as graduates of the École de foresterie were promoted to the title of forest agent.[61]

With a school the faculty of which was composed of several members of the Forest Service, Piché wanted to design a study program that corresponded to the policies that he wished to see adopted province-wide. The Forest Service trained the personnel that he needed to implement his administrative and scientific programs, knowing that graduates of the École de foresterie could find employment in the department or in private enterprise, notably among timber limit holders responsible for implementing the department's policies. In 1914, 18 of the 27 graduates of the school were working at the Department of Lands and Forests, and the other 9 were working elsewhere in the public administration or in companies.[62] Between 1910 and 1940, 66 of the 181 graduates from the school worked in the Department of Lands and Forests; some 20 others were hired by the Department of Colonization, which took over responsibility for soil classification after 1925. Lumber and paper companies employed 35 of them, and the majority of the 42 consulting engineers worked for companies or the Forest Service conducting the forest inventories required by a 1922 statute.[63]

By establishing the Forest Service, Piché intended to "create an independent corps inspired by a real spirit of rectitude, and who will be essential to compliance in the country and even by politicians."[64] It was his response to the premier's desire to re-establish public trust in "the administration of Lands and Forests." Of course, the École de foresterie formed a needed recruiting pool, but in Piché's eyes it was not solely a question of providing a specialized education. It was also crucial to "inculcate the same views and ideas regarding forestry issues in each of its members."[65] Piché regarded debates on limiting the diameter of cut timber, which had been the basis for governmental policies in recent decades, as "puerile discussions." He intended, rather, to base future forest policies on silvicultural methods for which only "educated employees could ... prescribe and direct application."[66] Reforestation, which he taught at the École de foresterie, was one of these new methods. It was integrated into a training program and into a forest policy through which the government sought to legitimize its reclamation of land abandoned to colonization.

Public Tree Nurseries and Private Plantations

Reforestation seemed to be both a knowledge tool and a technique for modelling the forest cover and proceeding with separation of the agricultural and forest areas. From its foundation, the École de foresterie initiated students to the methods and aims of reforestation. The program included theoretical and practical sessions on "artificial regeneration," and the students were required to take courses at the Pépinière provinciale de Berthierville, where they performed practical internships. With regard to technical work and courses, the school's general rules stipulated that "at the end of the year, they [the students] will have to present a report on the reforestation work that they will have done at the Berthierville tree nursery or elsewhere."[67] Every spring, the students did their apprenticeship at Berthierville and assisted government employees with sending plant materials and with seeding and transplantation tasks. During the summer, they participated in the Forest Service's technical and experimental activities, which included research programs on silvicultural systems and the cultivation of indigenous and exotic trees. This complementarity between research activities in the Forest Service and training at the École de foresterie was reproduced when the Rangers School was created in 1923 and the Station expérimentale in 1930, both situated at the provincial nursery.[68]

It was at the Berthierville nursery, on cleared land undergoing a desertification process, that the first lessons on reforestation took place, as did early work by the Forest Service in this area. Piché took inspiration from the French experience in Les Landes, where beachgrass was used to reforest abandoned land buried under drifting sand.[69] He believed that the creation of an artificial forest on deserts created by human activity could convince nearby rural populations of the possibility of returning infertile soil to production through the cultivation of trees.[70] The Forest Service did the same thing in Argenteuil County in the lower Laurentians between 1912 and 1925, when Avila Bédard recruited graduates of the school in early May to reforest the dunes at Lachute.[71]

The Forest Service conducted its reforestation activities directly on parcels of land stripped bare by fire or intensive cutting, but avoided the wide strips of forest in timber limits. Piché felt that due to the risk of fire the only reasonable action was to let nature "take care of bringing back the forest itself."[72] In his first annual report, he made his position clear: "We hear many people say that we should see to reforesting burned areas, and so on. I am not in favour of this work on a large scale. I am in favour of reforesting parcels stripped by exploitation or by fires only where we can be certain that fire will not ruin this work."[73] The allocation of resources to artificial forest regeneration was at risk as long as an effective fire protection organization was not in place. It was up to the timber limit holders to see to regeneration of the forest in their territories.[74]

Did this mean that the Department of Lands and Forests was turning its back on timber limits and the pulpwood that was cut in them? Although, with the approval of its personnel, the department was hesitant to undertake large-scale interventions, some paper companies reforested parts of their limits; among these, some maintained a nursery.[75] In fact, it was in response to requests made by company executives to the department that Piché and Bédard had conducted some research on reforestation during their studies at Yale. Unlike sawmills, which could easily be moved to be close to new timber stands, paper mills were permanently situated in their location. Considering the high fixed costs and the socio-economic infrastructure of paper mills, it seemed less onerous to these companies to plant a new forest when the resource was exhausted.

One example of private enterprise that explored reforestation of limits and maintenance of plantations was Laurentide Paper Company in Grand-Mère, in the Saint-Maurice Valley.[76] Between 1908 and 1912, the head of Laurentide's Forestry Division, Ellwood Wilson, conducted a series of studies in order to convince the company's executives to commit to a plantation and reforestation program in its limits. In Wilson's view, reforestation would be more profitable than intensive exploitation of distant forests. The best trees in the region had already been consumed by the sawmills, and the company had to look toward the northern parts of its limits to find wood to cut. Going to these territories and bringing the wood back was a challenge, as road infrastructure was either nonexistent or insufficient and floating wood on the Saint-Maurice River was a "long, hazardous, costly and onerous enterprise."[77] Wilson had tried to interest the Department of Lands and Forests in purchasing vacant land near Grand-Mère for reforestation, but Piché let it be known that his first commitment was to reforesting the area around Berthier County and that he hoped that these "experiments and projects will be attentively followed by the public and will create many more attempts of the same type by private individuals."[78] Wilson then prevailed upon Laurentide's executives to purchase the land, on which he created a nursery; he then engaged the company in a reforestation program between 1912 and 1931.[79]

Although he claimed that holders of timber limits alone were responsible for replanting their vast stretches of forest, Piché was not indifferent to the paper companies' reforestation efforts, and he tailored the Forest Service's activities to the needs of that industry. Through the provincial nursery, the Forest Service supplied the technicians and trees needed for production of pulpwood forest in Quebec. Demand from limit holders made white spruce the most sought-after seedlings at the Berthierville nursery (Table 3.1), even

TABLE 3.1 Distribution of seedlings at the Berthierville nursery

Species	1909	1916	1927
White spruce *(Picea glauca)*	1,500	7,150	2,613,056
Red pine *(Pinus resinosa)*	1,000	8,058	55,567
Norway spruce *(Picea abies)*	30,000	179,162	45,317
Scotch pine *(Pinus sylvestris)*	25,000	81,956	8,046
White pine *(Pinus strobus)*	75,000	196,552	20,073

though, in the early years, the plantations were composed mainly of ornamental trees and hardwood species, as well as white pine, which fell victim to blister rust in the 1910s. For the foresters, a key factor in the selection of the species cultivated in nurseries was speed of growth, as the main goal was to accelerate a process considered too slow to ensure the sustainability of lumbering operations.[80] The species also had to show other characteristics common to plant breeding, such as resistance to frost and disease. Above all, it had to correspond to the requirements of pulpwood production, whether it was grown in sand, on agricultural land, in the forest, or in the limits of lumber and paper companies. Recognized as good sawtimber, spruce was also appreciated for the quality of its wood fibre. It was mainly this last criterion that oriented production and distribution of seedlings at the provincial nursery.[81]

And so industrial demand, mainly from Laurentide but also from other pulp and paper companies (nine of which had reforested 40,500 acres of Crown land between 1908 and 1929, using 23.3 million trees and 143,000 pounds of seeds), oriented production and distribution of seedlings at Berthierville.[82] White spruce, in particular, was highly prized by Laurentide, which, since 1916, had expanded the capacity of its nursery by making annual orders from the provincial nursery. Wilson had initially obtained trees and seeds from a provincial nursery in Ontario to start up his operation, but he turned to Berthierville to organize Laurentide's nursery in Proulx in 1912. Situated a few kilometres north of the Grand-Mère plant, the nursery was to supply Laurentide's timber limits with some material grown on site and some purchased from Berthierville. This arrangement continued until 1924, when the Laurentide nursery became self-sufficient. Most of the 20 million trees that Laurentide planted between 1908 and 1930 were spruce.[83]

The influence of the paper industry on the Forest Service's plantation and reforestation projects became even more obvious when, during the Depression, companies abandoned their own activities in this area. In 1931, the Canadian Paper and Power Corporation (CPPC), which had absorbed

Laurentide several years earlier in the wake of a wave of consolidations in the paper industry, concluded an agreement with the Forest Service for maintenance of the Proulx nursery.[84] Facing financial difficulties and in arrears on stumpage fees, the CPPC was forced to give the nursery to the department as security. The agreement stipulated that the Forest Service would run the nursery for five years, after which, if the company failed to reimburse the maintenance costs, it would become state property.[85] In October 1931, the CPPC informed Wilson that no money could be devoted to reforestation and that his services were no longer needed.[86] Clearly, the company did not plan to continue its forester's work, and Wilson, disappointed, left for the New York State College of Forestry.[87] On February 28, 1936, the CPPC definitively abandoned the nursery.

Although the department and private enterprise seemed to have different purposes for their reforestation strategies – the former, to arrest desertification on abandoned private land, and the latter, to rebuild wood resources on leased Crown lands – the activities conducted at their respective nurseries remained, on the whole, identical. A look at the fate of the Proulx nursery brings to light how similar the private and public reforestation efforts were. That the Forest Service had the technical skills to plant 20 million white spruce trees indicates how closely its previous commitments corresponded to the reforestation strategies of the pulp and paper industry. But where did these seedlings end up? And how did reforestation fit within the department's policies?

Production of a Pulpwood Forest

As the hub for a network of institutions behind implementation of scientific forestry in Quebec, the Berthierville nursery became a lever for reclaiming land lost to colonization. The acreage acquired by the Department of Lands and Forests for its nursery in 1907 was a farm that had been abandoned six years earlier. The farm's soil, invaded by sand, was an example of a common phenomenon at the time, desertification, the result of negligent practices by settlers who had ruined their land. On one thirty-acre section of the lot, Piché and Bédard planted trees to be used for reforestation, such as pines and larches, as well as exotic species that they hoped to acclimatize. A second section, measuring forty acres, was used for agriculture, with cultivation of cereals and potatoes. In his first annual report, Piché emphasized the harmonious complementarity of agricultural and lumbering activities by giving a full description of work on the farm and presenting data on crop yields, rotation and fertilization techniques, and soil fertility.[88] Piché felt that the farm functioned as a demonstration site and would encourage farmers to reforest part of their infertile land with a woodlot. However, as the area devoted to the nursery grew, in part to respond to growing demand by paper companies, the coexistence of the agricultural and lumbering activities was compromised. The space needed for seeding and transplanting impinged on the area devoted to agriculture, and, in 1918, the Forest Service terminated its agricultural activities in Berthierville after successive expansions of the nursery and construction of a seed dryer.[89] By this time, the Department of Agriculture was already operating its own nursery at Deschambault, as the facilities at Berthierville did not correspond to the needs of the fruit-growing industry that it was trying to establish (see Chapter 5).

Providing the basis for the scientific practices of the Forest Service and the training at the École de foresterie, the process of reclaiming the land in Berthierville that had been abandoned and bought up by the Forest Service seemed to be a microcosm of reforestation activities province-wide. As mentioned above, the Forest Service pursued its reforestation activities on parcels of land laid bare by fire or intensive cutting; the aim was to make soil that had become barren productive again and to educate people about the importance of forest conservation. Also, Piché refused to engage in a large-scale reforestation program in timber limits until companies instituted a fire protection system. In 1925, he recognized that little progress had been made in this regard outside of the desertified land over the previous ten years.[90]

The service nevertheless undertook an intensive reforestation program in vacant lots in rural municipalities and township reserves after the number and effectiveness of firefighting organizations grew, leading Piché to declare that "the fire problem is, so to speak, resolved."[91] It must also be said that the Forest Service, which had just lost surveillance of timber cutting on lands to the Department of Colonization, could gain some control over settlers' lumbering activities through the intensification of forest management operations in the township reserves. In 1925, at the request of Piché, who wished to implement a reforestation program on more than 2 million acres of land, the Legislative Assembly of Quebec voted an annual disbursement of $100,000 to "promote, assist, and encourage reforestation work by the gathering of seeds, the maintenance of nurseries, the acquisition of land, reforestation, distribution of seedlings, and the plantation of forest and ornamental trees in the province."[92]

The first entities targeted by the reforestation program were the municipalities, which, in Piché's view, were responsible for the fires that started in their territories and then spread to the timber limits.[93] Debris from cutting was strewn across the ground of abandoned woodlots that were either not reclaimed or not subjected to supervision to limit the extent of the fires that started in them. In 1922, the government adopted a statute that allowed municipalities to acquire lots abandoned by their owners in exchange for unpaid taxes. Although the objective of fighting fires was reiterated, fiscal considerations also encouraged the government to promulgate this law, through which it hoped to transform "abandoned land under the responsibility of municipalities due to unpaid taxes into forests that generate revenue and constitute an important supplementary natural asset."[94] However, the municipal forests that the Forest Service was hoping to create had to contribute specifically to industrial development. "These lots could supply more than one million cords of pulpwood every year if they were intelligently managed and cultivated – that is, as much as Canadian plants currently consume and as much as we export to the United States each year."[95] To make this happen, Piché proposed to repeat the strategy employed for reclamation of the dunes in Lachute: have the department purchase the razed and denuded land in order to reforest it and sell it back to its original owner. The municipality might then "obtain perfect title to these lots and use them to form the nucleus of a communal or urban forest."[96] If a municipality refused to participate in the program, the Department of Lands and Forests would take possession of the land, reforest it, and integrate it into the township reserve system. This was in fact the fate that Piché chose for the Lachute dunes when the original owners refused to repurchase their completely reforested land, as he had intended in 1912.[97]

In the end, Piché had to concentrate the Forest Service's reforestation efforts in the township reserves; he was unable to obtain the cooperation of the municipalities, which were utterly uninterested in maintaining communal forests. After the first township forest reserves were created, in 1911, they continued to spread across the province, the growth in their numbers and area (see Figure 3.3) and their locations (see Figure 3.4) driven by the colonization goals of Quebec governments. Until the mid-1920s (the service had seventeen reserves covering 265,000 acres in 1922), the vast majority of these reserves were situated in Lac-Saint-Jean, Roberval, Chicoutimi, and Saguenay counties. Also in this region was the enormous forest reserve of the Laurentide National Park. The park bordered colonization zones in the Saint-Maurice and Lac-Saint-Jean counties, north of Quebec City, where, as geographer Pierre Biays noted, "These [township forest] reserves constituted a rampart against expansion of the ecumene. Revoked sold lots, authorized repurchased lots, and transferred Crown lands were integrated into reserves certainly designed to supply wood to *habitants* and settlers, but mainly to stop the leasing of lots."[98]

FIGURE 3.3 Number and area of township forest reserves in Quebec, 1911–39

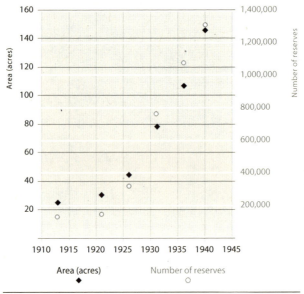

Source: RAMTF, 1911–40.

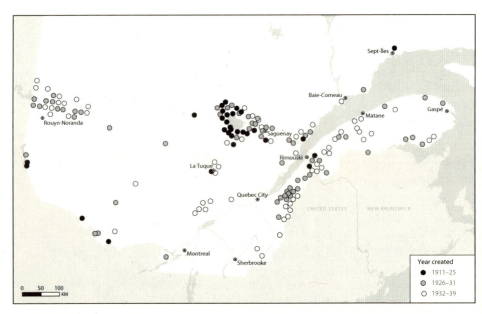

FIGURE 3.4 Distribution of township forest reserves in Quebec, 1911–39. | *Source: RAMTF, 1911–40.*

The Department of Lands and Forests began to create township reserves in the frontier district of Abitibi and in the Lower St. Lawrence in the late 1920s to support colonization in these regions. Most of these reserves were in the counties of Matapédia and Témiscouata, as well as those of Montmagny and L'Islet; at the same time, temporary nurseries were created to accelerate forest regeneration. By giving settlers access to reserves of standing timber, the Department of Lands and Forests was seeking to protect the large areas that it was reserving for the future of the logging industry. In the 1930s, the service established township reserves in all regions, although most were situated in counties in the Lower St. Lawrence and Abitibi, where the majority of settlers were sent under the Vautrin plan during the Depression.[99] By 1938, there were 122 of these reserves, covering 942,000 acres. Although it had created them to support settlers' need for wood, the Department of Lands and Forests quickly came to appreciate their future role in occupation of the territory. As their number grew, the department foresaw that they could supply material needed for the emergence of small industries and create jobs for farmers outside the growing season. In 1922, forest rangers began to plant spruce and pine in the forest reserves in the Parke and Kénogami townships "to compensate for the slow rate of natural regeneration."[100] Despite continuing illegal cutting in both the timber limits and the reserves, the township forest regime enabled the Department of Lands and Forests to protect great stretches of forest and the reforestation work that it was doing in them. It sanctioned this approach by classifying most township forests as permanent forest reserves. The Forest Service then extended its reforestation program beyond simply reclaiming desertified land and vacant lots, at the same time that the provincial nursery grew into a network of smaller nurseries to supply the township reserves with seedlings adapted to local climatic conditions. Starting in 1930, the secondary nurseries at Macpès, Amos, and Normandin served the regional needs of the ten township reserves that were slated for reforestation: those in Lachute, Jonquière, Normandin, Macpès, Parent, Albanel, Caron, Cimon, Parke, Kénogami.[101]

During the 1930s, the Forest Service sought to ensure the long-term availability of the resource by including the reforestation of wide swaths of forest in its mandate, and it stepped in for companies that abandoned their reforestation efforts in the wake of the economic crisis. At the beginning of the decade, Piché announced: "We propose to be able to plant at least fifty million trees per year within ten years, in order to accomplish the program that Mr. Mercier, Minister of Lands and Forests, set out for us when he asked us to organize to plant at least twice as many trees as are felled each year."[102] Although the Forest Service had to cut back on reforestation activities and distribution of seedlings from the provincial nursery following a drop

FIGURE 3.5 Originally a township reserve provided for settlers to collect wood, the Parke Reserve was home to a Department of Lands and Forests temporary nursery to supply plants to other township reserves in the process of reforestation. | Nursery in the Parke Reserve, by J.W. Michaud, 1942. *Source:* BANQ-Québec, Fonds ministère de la Culture et des Communications, Office du film du Québec, Documents iconographiques, E6, S7, SS1, P7882.

in demand by timber limit holders in 1931, it nevertheless accelerated reforestation of township reserves after acquiring the Proulx nursery that year. Even though the agreement with the CPPC provided that the company would take back the nursery after reimbursing the maintenance costs, the Forest Service wished to avoid the loss of more than 12 million seedlings there. It therefore transferred them to its secondary nurseries or planted them in its reserves. In 1938, the network comprised eight nurseries, situated in Berthierville, Proulx, Trécesson, Normandin, Roberval, Parke (see Figure 3.5), Macpès, and Ouimet. At these facilities, the service maintained a stock of young trees for reforestation, with total production varying between 4 and 5 million seedlings. The provincial nursery in Berthierville supplied these nurseries with seeds after construction of a granary for processing and extraction of thousands of cones. The entire network helped to reforest the growing number of township reserves, in which the service planted 21,081,206 trees, mainly white spruce from Proulx or Berthierville, and sowed 44,964 pounds of seeds on about 50,000 acres between 1925 and 1938.[103] Although the Forest Service acknowledged at the end of the 1930s that it had been unable to reach

FIGURE 3.6 Like all large nurseries, whether public like Berthierville or private like the Laurentide nursery in Proulx in the Saint-Maurice Valley, reforestation and nursery activities in the township reserves were focused on species that could supply the paper industry. | Pile of pulpwood in the Parke Reserve, by J.W. Michaud, 1942. *Source:* BANQ-Québec, Fonds ministère de la Culture et des Communications, Office du film du Québec, Documents iconographiques, E6, S7, SS1, P7872.

the reforestation objectives set by Piché at the beginning of the decade, it nevertheless profited from the Depression to perpetuate a reforestation policy targeting the creation of pulpwood forests (see Figure 3.6) – forests growing on land developed to regulate the colonization movement throughout the province.

AS THE CONFLICT BETWEEN promoters of colonization and defenders of the forest industry grew and each faction sought to expand its grasp on the hinterland, the protection of wood resources and access to Crown lands became a central issue for the public finances of the state and for industrialization of the province. Soil classification and the division of land into spaces for colonization and for forest exploitation appeared to be the political and technoscientific solution to the conflict.

Left to itself, the forest could not regenerate quickly enough or respond to the demand of the quickly growing pulp and paper industry that the Liberal government was trying to support. In the early twentieth century, when the Department of Lands and Forests created its scientific apparatus and training site to govern forest exploitation, it was seeking more to encourage the economic development of the province than to respond directly to the demands of the industry, which, through its own nurseries and reforestation projects, was ensuring the regeneration of its timber limits. It was important for the department to create pulpwood forests and resolve the opposition between the colonization movement and lumbermen. When the companies abandoned their reforestation efforts and nurseries in 1930, the department could take over these activities precisely because its previous investments – its infrastructure, its policies, and the personnel in its Forest Service – fell within a mission that included the reforestation of pulpwood forests, which was what the companies had been doing. At the same time, it was able to calm the tempers and demands of colonization advocates by creating forest spaces that met the need for wood of both settlers and paper mills.

The government of forest resources was based on a series of measures involving the separation of forestland and farmland through soil classification, the shaping of the landscape through reforestation, and the regulation of access to the forest and its resource by placing parcels of Crown lands in reserves. To do this, the Department of Lands and Forests made use of scientific forestry – not so much for its theories as for the personnel and techniques that would enable it first to survey the territory in order to describe and delimit it, then to govern its human and ecological entities. The allocation of township reserves throughout the province, the distribution of reforestation materials, and the underlying network of nurseries contributed to the separation of forestland and farmland and to the protection of the forest and its resources for the benefit of the paper industry. Thus, a central practice of scientific forestry, reforestation, made it possible to renew the prospects for the forest space administered and shaped by the Department of Lands and Forests while complying with the social and economic policies of Quebec governments.

4
Surveillance and Improvement of Fish and Game Territories: Conservation of Wildlife Resources

LIKE THE FOREST, wildlife was regulated by measures based on creating reserves and shaping the landscape. Unlike measures for the forest, however, those for wildlife were instituted mainly through a devolution of responsibilities from the public administration to private entities. Through their fish and game clubs, members of the North American economic and political elite leased territories from the province. Because the leases included the obligation to protect fish and game over the long term, the clubs made improvements to their leaseholds and hired wardens to enforce the laws and regulations on hunting and fishing in order to ensure the perpetuation of faunal populations for exploitation.

Provincial departments responsible for fish and game were unable to assess the viability of terrestrial and aquatic wildlife except in general terms. In 1883, the commissioner of Crown lands appointed a superintendent of fisheries to inspect the salmon rivers in the Gaspé peninsula, the Côte-Nord region, and the Saint-Maurice Valley to determine their "ichthyological value" according to the abundance and size of specimens and the number of "fishing spots." Apart from a report on wildlife in the Côte-Nord written by inspector general of fisheries and wildlife Henri de Puyjalon, who was sent as a special envoy to Labrador in the late nineteenth century, no systematic wildlife survey was conducted in any region before the 1930s.[1] The superintendent therefore depended on sportsmen's perceptions to assess the availability of fish and game. The private clubs were required by law to provide information regarding results of their catches for "the proper administration of the fishing business in this province," but few of them lived up to this obligation. The superintendent had to constantly remind them of their commitment, and he regretted that "the statistics [that the lake anglers are supposed to supply to the department regarding the quantity of fish that they catch] remain incomplete ... thus, through this negligence, we are unable to establish the real wealth of our lakes."[2] For the superintendent, it was important "to make the value of our fish and game resources better known and appreciated" so that the highest possible price could be obtained when parcels of land were put up for auction.[3]

The Quebec state shaped very early, and in various ways, the province's fish and game territory, even though its technoscientific personnel participated in the government of wildlife resources late in the period under study. In this chapter, I describe the conditions for the formation, definition, and use of this territory, as well as its supervision and improvement. First, land was leased and allocated to clubs under the statutes and regulations adopted by the Province of Canada. Then, the commissioner of Crown lands of the province of Quebec set up a surveillance service and created public reserves to protect fish and game. Wardens were sent out to control the subsistence practices and commercial activities of wildlife users who contravened the code of conduct of the sporting elite. Finally, using the hatcheries established under the administration of the Province of Canada and transferred to the provinces after Confederation, the Quebec state engaged in pisciculture to stock rivers and lakes and maintain the province's reputation as a prized tourism destination – "the sportsmen's paradise."[4] Indeed, the goal of governmental fish and game managers was precisely to keep the ichthyic and cynegetic resources abundant in an Edenic landscape in order to attract tourists, who spent a fortune in the province.[5] Because tourism appeal was based on plentiful wildlife, the government diversified its interventions by reinforcing its capacity to improve the territory for fish angling and game hunting.

Leasing Fish and Game Territories: The Private Clubs

The Quebec state was slow to intervene with its own agents to shape the fish and game territory because, since the mid-nineteenth century, private parties had been invested with powers to protect and exploit wildlife. In fact, when the Province of Canada leased salmon rivers in the Gaspé Peninsula to individuals in 1859, it recognized that it lacked the means to oversee such a large territory and regulate its use. It therefore chose to delegate these responsibilities to private clubs whose members committed themselves to protecting wildlife. This strategy spread throughout the province soon after

Confederation. By 1914, almost 300 fish and game clubs shared privileged access to the province's wildlife resources; there were more than 600 clubs in 1928, just before the Depression took hold, and 572 in 1939.

The granting of leases and the sale of hunting and fishing licences generated annual revenues that grew as the number of clubs increased (see Figure 4.1). What is more, an entire tourism industry developed around attracting sport hunters and fish anglers from large cities not only in Quebec but also in Ontario and the United States. In addition to stimulating local economies in remote communities thanks to the creation of jobs and circulation of currencies, that industry extended the ecumene beyond the St. Lawrence Valley.[6] The department responsible for the sector made minimal investments, however; basically, it covered the salaries of the superintendent and the inspector of fisheries and game – practically the only civil servants working in the sector until the First World War – and of the seasonally employed wardens, many of whom received low pay and some of whom were actually volunteers. Other state interventions, such as pisciculture and the creation of wildlife reserves, were self-financing thanks to user fees.

One of the explanations put forth by the Quebec government for its minimalist approach despite its interest in raising revenue and developing tourism was that wildlife conservation would require the mobilization of an

FIGURE 4.1 Quebec public revenues from fish and game club activities (constant dollars 1935–39 = 100)

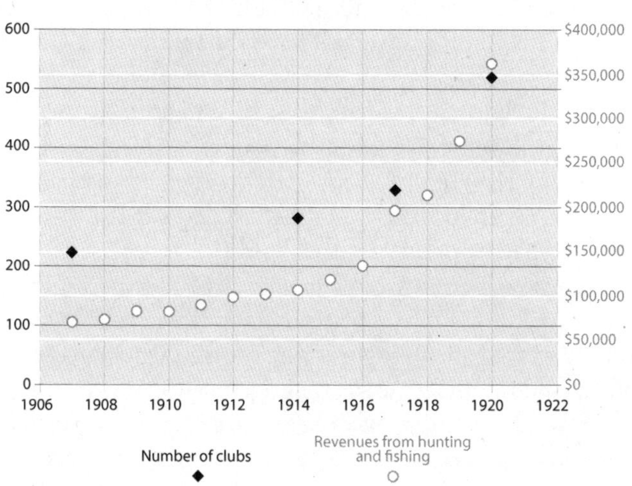

Source: *Rapport annuel du ministère de la Colonisation, des Mines et des Pêcheries (RAMCMP)*, respective years.

"army of civil servants" to cover a territory that was even more extensive because animals – unlike trees – were mobile, making the borders of the territory to defend and protect just as mobile. A similar situation existed for the improvement of wildlife habitats and their non-human inhabitants.

Pressured to protect fish and game over a vast territory – given the anticipated benefits for the public treasury – but lacking the means to do so, the government assembled portions of forest, lake, and river territories into tracts of various sizes. These tracts, which were on Crown lands, were leased out to private fish and game clubs. Having obtained exclusive use of vast stretches of land, these clubs paid handsomely to improve them for the pleasure of their members, usually numbering several hundred. Club members became the "wise stewards" of wildlife resources and Crown lands through their work and their investments.[7] In addition to ecologically transforming these wilderness areas, the clubs became agents of cultural change, as their members hired Indigenous and Euro-Canadian local residents as guides, in order to liberate them from their subsistence activities.[8] These jobs, they claimed, thus stopped the local people from using practices that put the viability of ecosystems and aquatic and terrestrial wildlife at risk.[9]

A number of historians have written about the origin of these clubs in Quebec.[10] The first to be founded was the Montreal Fish and Game Protection Club in 1859, one year after the Lower Canada Fishery Act was amended.[11] The conservationist ideas that were disrupting the forestry sector were also held by groups concerned about destruction of the rivers and the repercussions for aquatic fauna.[12] Their first target was protection of overfished rivers in the maritime estuary and the Gulf of St. Lawrence. Since 1852, a representative of the colonial state, Pierre-Étienne Fortin, designated as a stipendiary magistrate, had to travel hundreds of kilometres of shoreline to ensure compliance with the regulations and see to protection of the fisheries in the Gulf of St. Lawrence.[13] Given the failure of this approach, the legislature amended the law in 1858 to institute a system of leases and licences to regulate access to salmon rivers and sportfishing. The rivers were subjected to a system of riparian rights and leasehold tenure under which only rod-and-line fishing was permitted, while commercial activities in the estuary continued in the form of fishing stations under licence. Commercial fishers had to comply with regulations governing the fishing season and equipment so that fish would be able to access their upstream spawning grounds.

The system was aimed at stimulating investments by individuals in surveillance and improvement of watercourses – interventions that would otherwise have been provided by public funds, which were unavailable.[14] In exchange for paying rent and agreeing to comply with the rules respecting, among other things, protection of spawning grounds, club members acquired exclusive angling rights on salmon rivers, two of which, the Matapédia and the Ristigouche, were situated on Indigenous land.[15] The tenure system subsequently extended to the Cascapédia and Bonaventure Rivers, in the Gaspé Peninsula, and to the Godbout, Moisie, and Natashquan Rivers in Saguenay County, on the north shore of the Gulf of St. Lawrence. Anglers obtained exclusive rights over a half-mile of frontage on either side of the river for five, then nine, years, accompanied by restrictions on the length of the fishing season, capture methods, and quotas and the obligation to have wardens monitor the territory under lease and the activities conducted there. The fishing clubs committed themselves to hiring both wardens and detectives, who would arrest and bring suit against anyone committing infractions that compromised the survival of the resources.

In the decades that followed Confederation, Canadian and American sportsmen's private clubs became stewards of more than thirty salmon rivers in Quebec. In addition to hiring wardens, the clubs shaped the riparian environment by stocking the rivers with fish, dredging them, and building dams to regulate flows and increase the productivity of fish habitats. They received assistance from the federal Department of Marine and Fisheries, which, in 1873, established hatcheries in Tadoussac, Gaspé, and Restigouche to accelerate the stocking of rivers, whether or not they were leased out. Millions of fry were released into salmon rivers on the north shore of the Gulf of St. Lawrence and in the Gaspé Peninsula.

Intervention by the federal administration in the management of fish resources following enactment of the British North America Act is explained by the ambiguous constitutional status of watercourses situated on Crown lands that had just been transferred to the provinces. The fate of these watercourses was not clearly settled; navigability was the criterion for whether they fell under federal or provincial jurisdiction.[16] The stakes were high, because under the law in effect in Lower Canada, and then in the province of Quebec, the holder of a property bordering a river could rent or sell its use to a third party separately. On 28 April 1882, a Privy Council Order acknowledged that the federal government had to cede both its rights over inland waters situated on Crown lands to the provinces and the fishing rights granted to riparian landowners.[17] In March 1883, the Quebec commissioner of Crown lands undertook to grant, by lease, exclusive fishing rights on the shores of parcels of riverine land. As they had been under the federal system previously in force, tenure holders remained responsible for application of the fishery laws. They obtained the right not only to bring suit against offenders but also potentially to have the costs of trials reimbursed. Finally, in 1884, the Quebec government, inspired by the French civil code, applied the "three chains" legal reserve, a provision that recognized a corridor of almost sixty metres in depth on land bordering rivers and lakes to be reserved for fishing.[18] By doing this, it appropriated exclusive power to lease the fishing rights to all of the province's inland waters.[19] Later, the fish protection system was expanded across the province, beyond the estuary and tributaries of the St. Lawrence.

Game territories were subjected to a tenure system similar to that instituted for the salmon rivers of the St. Lawrence estuary. Unlike jurisdictions in the rest of North America, the Quebec state leased land to private clubs, even though it adopted identical regulations regarding the length of the hunting season, the age and sex of the animals hunted, the practices and equipment, and licences and quotas.[20] Like aquatic fauna and the practice of angling, terrestrial fauna was subjected to a protection strategy favouring sport hunting over all other forms of capture, such as commercial and subsistence hunting. The commissioner of Crown lands delegated responsibility for surveillance of the cynegetic territory and implementation of hunting regulations to clubs. The clubs' wardens were invested with state authority to enforce the law, as were the commissioner's few wardens, hired seasonally starting in 1876. As well as surveillance of the fishing and hunting territory by private individuals and production of revenue for the public purse, the system of private clubs encouraged improvement of the territory by non-state actors, notably through construction of lodges and roads.[21]

In the view of historian Darcy Ingram, it was the creation of Laurentide National Park that caused the leasing system for angling rivers to be extended to hunting territories.[22] As noted in Chapter 3, the commissioner of Crown lands created the park in 1895 to protect forest, hydraulic, and wildlife resources. However, this "reserve" threatened the activities of certain fish and game clubs that had leased tracts now situated within the park. Although they had an exemption from park regulations for fishing, their hunting rights were not formally recognized. Members of the Triton Club, which was in this situation, met with the commissioner to obtain hunting rights on the territories that they were leasing inside the park. The commissioner saw this as an opportunity to lease hunting territories in order to generate revenues and establish a wildlife protection system. Although some members of the Legislative Assembly were opposed to granting foreigners privileges to the detriment of local inhabitants and demanded the right for settlers to hunt on these territories, the Assembly adopted, on December 21, 1895, a regulation granting the Triton Club hunting, fishing, and shooting rights on the leased portions of land situated within the park as of January 1, 1896.[23] The lease freed club members from the obligation to acquire hunting licences for a ten-year period. In 1896, 24 more clubs signed leases on territories with a total area of 3,042 square kilometres. By 1900, 78 clubs had hunting rights in areas totalling 9,103 square kilometres; 342 leases for rivers, lakes, and hunting divided up the province's wildlife territory among 113 fish and game clubs. There were 283 clubs in 1914, which had signed 557 leases, including 189 for hunting territories covering over 23,200 square kilometres.[24]

In the final years of the nineteenth century, private fish and game clubs proliferated throughout the province, and territories under lease dotted both shores of the St. Lawrence, from the Ottawa River to the Gaspé Peninsula. With extension of the railway, the clubs gained access to remote sites north of the Saint-Maurice Valley and Lake St. John and further west, in the Témiscamingue region.[25] Some clubs were formed around small groups of wealthy individuals who mobilized their economic, financial, and industrial networks and combined leases to increase the area of their hunting and fishing territories and form large domains. These included the Shawinigan Club (1883), the St. Bernard Fish and Game Club (1872), and the Winchester Club (1880) in the Saint-Maurice Valley, as well as the Restigouche Salmon Club (1880) and the Club de chasse et de pêche de Rimouski (before 1885) in the eastern part of the province.

As the sole interlocutor for sportsmen, the Quebec government helped things along by adopting the Act to Facilitate the Formation of "Fish and Game Protection Clubs" in the Province in May 1885.[26] The statute encouraged fish and game clubs to incorporate so that they would become financially and legally responsible for conservation of wildlife resources, whether by improving their tenured tracts or by prosecuting those who violated the hunting and fishing regulations. After the statute was enacted, other prestigious clubs were founded, including the Laurentides (1886) and Triton (1886) clubs, north of Trois-Rivières and Quebec City, respectively, and the Mégantic club (1888), astride the Canada-US border, in the Eastern Townships and Maine. The number of members and the area leased varied from club to club, as the statute established no specific criteria to this effect until an order-in-council in 1901 limited the tenures to 200 square miles. As it was harder and harder to find land, especially around the Laurentian and Appalachian ranges, and in the Gaspé to the east and Pontiac County to the west, clubs created later had to be content with medium-sized tracts containing a few lakes.[27]

The government and sportsmen therefore maintained a privileged and mutually beneficial relationship following adoption of the 1885 statute. On the one hand, the government filled its coffers and was guaranteed the implementation of a policy of wise stewardship of rivers and Crown lands; on the other hand, sportsmen obtained exclusive access to coveted destinations and at the same time saw to protection of wildlife resources, serving their own interests. The mutual character of this relationship was clearly manifested in the title – and the content – of a 1914 government publication, *The Fish and Game Clubs of the Province of Quebec: What They Mean to the Province, What Privileges They Enjoy.*[28] This official publication strongly defended the system that had been in place for more than thirty years. First, it recognized that the Quebec government and the province's economy received large amounts of money from sportfishing and hunting, of course through leasing fees and the sale of permits but also from expenditures made by tourists in local communities. The author noted that annual revenues from this activity were equivalent to $2 million and surpassed what the commercial fisheries had generated in 1911. What is more, the expenditures made by the clubs to develop their tracts and to monitor wildlife and protect it against overexploitation lightened the burden on the public treasury, as the government did not have the resources to establish an effective warden system for the protection and surveillance of a huge territory. In the view of the department responsible for fish and game, such a function had become essential to avoid depletion of wildlife resources, without which the sector would lose its attractiveness to foreign sportsmen. The author brought up the contemporary cases of extinction of the passenger pigeon and the problematic situation of bison, which were threatened with extinction, to praise the protective role of the private clubs in the preservation of wildlife.[29]

Surveillance of Wildlife and Hunters: Guarding Public Lands

The creation of fish and game territories could not in itself ensure protection of wildlife without adequate surveillance of the territory, and so the hiring of wardens by private clubs was just as crucial as leasing of the territory to avoid depletion of wildlife resources. The state therefore invested the wardens with the legal authority to arrest people who contravened the regulations regarding capture of fish and game and to bring suit against the offenders. It should be noted that, well before the leasing of hunting and fishing territories and the setting up of private clubs, the state had its own guardians to oversee the fisheries. Indeed, after adoption of the Act for the Protection of Fisheries in Lower Canada in 1855, fishery protection vessels were sent to patrol the estuary of the St. Lawrence and the rivers along the north shore and in the Gaspé to prevent depletion of salmon rivers. As it had only twenty-three civil servants to cover all of the waters in the province of Quebec, the federal government decided in 1869 to encourage private clubs established on the Moisie, Natashquan, Watshishou, Cap Whittle, and Anticosti Rivers to hire their own wardens to ensure application of the fisheries statute. Federal civil servants received an average of $30 per year to monitor maritime fisheries and watercourses on unleased Crown land on the north shore of the St. Lawrence and in the Gaspé Peninsula, so the clubs used the federal government's compensation to enhance the payments that they made to their wardens, whose salaries ranged from $50 to $250 per year.[30]

Soon after the Privy Council Order of 1882 ceded full responsibility for inland fisheries to the provinces, the Quebec commissioner of Crown lands intensified the surveillance instituted under the federal regime. From eleven in 1875, the number of fish and game wardens grew to eighty-two in 1889, six years after the Fisheries Service was created. The first Quebec superintendent of fisheries, C.D. Mackendie, suggested in a report, however, that the priority of the surveillance system was not so much to prevent overexploitation of wildlife as to "increase the value of rivers, leased or not," as the price of the leases could then be negotiated upward when they were auctioned off.[31] J.-N. Proulx, who succeeded Mackendie in 1885 after a year in the position of superintendent of game, reiterated the need for "those who lease lakes or rivers, in compliance with this clause of their lease ... to recommend the official appointment of a fishing warden for the waters under lease in their favour," again with the goal of increasing the "ichthyological value" of watercourses and "creating greater competition on the day of the auction."[32]

The superintendent continued to support the surveillance efforts undertaken by private clubs in ensuing decades. In addition, he constantly expressed his concerns about the working conditions for public wardens and the effectiveness of the surveillance system itself. Given the size of the areas that the wardens had to cover, the low pay they received, and their close relations with surrounding communities, the superintendent wanted to increase their wages. In 1893, superintendent H. Chassé hoped that "sufficient pay for game wardens" would help to prevent "the disappearance of game and fur-bearing animals." A committee formed by the government recommended "that the game service be organized on a better scale," which would entail making "new appointments and revoking the commissions of most of the old game wardens, who, for a number of reasons, are not fulfilling their job requirements."[33] During his short tenure as superintendent of fisheries and game, Louis-Zéphirin Joncas established two priorities for wildlife protection: reducing the number of wardens and raising their pay, and creating small-size fish and game reserves that could be monitored by fewer wardens, "particularly in the Métapédia [sic] Valley, where poaching is taking place on a large scale."[34] Joncas's successor, Hector Caron, reiterated the need for fewer and better-paid employees, as, in his view, "we cannot reasonably expect that these men will ensure that the law is obeyed."[35] For example, as the government was not giving in to repeated demands for the creation of public reserves in the Matapédia Valley, Caron recommended that the Fisheries and Game Service employ "ten men who are skilled, energetic, and sufficiently paid that they can give all of their time to the duties of their responsibilities" and that the law be amended "to extend the statute of limitations for an offence and therefore facilitate successful lawsuits."[36]

In 1910, the Fisheries and Game Service changed its wardens' working conditions: it reduced their numbers considerably (see Figure 4.2), from 205 to 132, and raised their salaries from $18,000 to $31,865 (in constant dollars). Subsequently, the number of wardens increased regularly until the mid-1920s, and their salaries also increased; a constantly expanding share of the Fisheries and Game Service budget was allocated to surveillance of territories. The number of public wardens began to drop in 1926 as the private clubs took over and hired more of them (see Figure 4.3). In the early 1930s, as the number of public wardens continued to fall, surveillance seems to have been relaxed somewhat. The minister announced that due to financial difficulties engendered by the economic crisis, wardens would adopt a permissive approach to accommodate local subsistence practices. With the Depression and the back-to-the-land movement, colonization territories became vulnerable to poaching, and the department did not dare to intervene given the widespread poverty.[37] Although both hirings and fine collections were on a downward trend in the 1930s, in the latter case the trend had begun a decade earlier (see Figure 4.4).

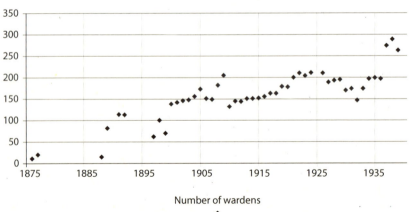

FIGURE 4.2 Number of wardens hired by the Fisheries and Game Service

Sources: Service de chasse et pêche (see note 5 of this chapter), *Rapport annuel*, respective years; *État des comptes publics pour la province de Québec*, respective years.

FIGURE 4.3 Number of wardens employed by the Fisheries and Game Service and by private clubs

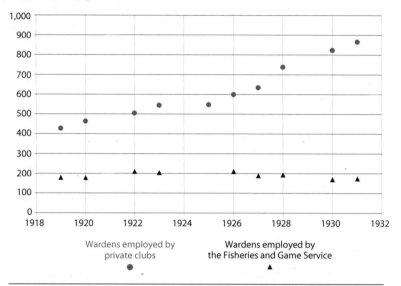

Sources: RAMCMP, 1918–31; État des comptes publics pour la province de Québec, respective years.

FIGURE 4.4 Amounts of fines collected (constant dollars 1935–39 = 100)

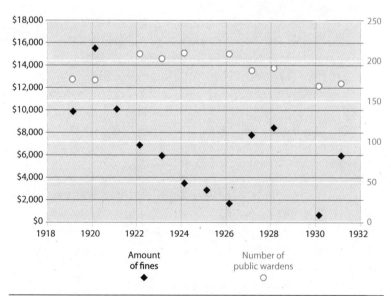

Sources: RAMCMP, 1918–31; État des comptes publics pour la province de Québec, respective years.

As the number of public wardens hired was falling, the superintendent of fisheries and game reorganized the service to have, "from one end of the province to the other, a corps of vigilant game wardens who will do their best, under conditions that are not always easy, to make the hunting and fishing laws known and obeyed."[38] A large number of wardens were posted to the Abitibi region and the Ottawa Valley, but other areas, such as Bonaventure and Gaspé counties in the Gaspé Peninsula, saw a substantial drop in staffing (see Figure 4.5). The service also assigned fewer wardens to Quebec, Portneuf, and Saguenay counties, but the trend had been underway in this district (Quebec–Lac-Saint-Jean) since the early 1920s, a few years after the massive hirings of fish and game wardens (see Figure 4.6). In the centre of this district was Laurentide National Park, which changed vocation to become a wildlife sanctuary in 1923, when the service knew that it could count on private clubs for surveillance of the territory (see next section). Notably, as the number of wardens employed by the Fisheries and Game Service dropped, a greater number of wardens were working for the private clubs, even though these clubs were becoming fewer and their respective domains were increasing in size. It was only in the late 1930s that the number of wardens working for the government began to increase again, mainly to protect areas heavily exploited since the massive influx of settlers and the back-to-the-land movement occasioned by the economic crisis, such as the Gaspé Peninsula, the Lower St. Lawrence, and the northwestern part of the province (see Figure 4.5). Nevertheless, a study published in 1941 showed that the state was still a relatively minor actor in wildlife protection: "In the clubs' territories, there is one warden per 16 square miles, whereas in the public domain a warden must monitor an area of 700 square miles."[39]

Historians have noted that in the early decades of the twentieth century, the leasing of much of the territory and surveillance by wardens employed by private clubs and the public administration seem to have suppressed poaching.[40] Does this mean that the warden system was effective or that wardens were reporting fewer offenders? In regions frequented by sportsmen, the wardens' reports show a high degree of both surveillance and contempt for the law. Rather than leading to a decrease in reprehensible activities, the implementation of increasingly severe regulations only highlighted the extent of illegal practices. The minister occasionally mentioned arrests of poachers, but few were actually charged. The wardens' effectiveness was even more questionable because their presence seems simply to have encouraged the poachers to move to areas that were not as well monitored.[41]

Aside from promotion and conservation of wildlife, one of the apparent objectives of the surveillance system was to persuade the resident population to adopt new fish angling and hunting practices, so that the province's

FIGURE 4.5 Distribution of game wardens and fish wardens by district

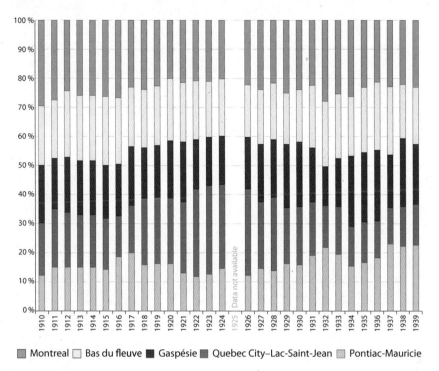

Note: The designation of districts corresponds to the residence locations of the inspectors attached to the Gaspésie (Matane), the south shore of the St. Lawrence River from the Eastern Townships to the Lower St. Lawrence (L'Islet), Montreal and surrounding counties (Montreal), the territory included between the Ottawa Valley and the western shore of the Saint-Maurice River (Shawinigan), and the territory east of the Saint-Maurice River to the Côte-Nord region (Saint-Jean, Île d'Orléans). These districts are listed in the 1916 annual report.
Sources: Service de chasse et pêche, *Rapport annuel,* respective years; *État des comptes publics pour la province de Québec,* respective years.

wildlife would become exclusive resources for sporting activities.[42] In his annual report, the minister underlined that "we must educate these people by demonstrating that protection of fish and game is definitively in the public interest, for the good of the community in general, and it even profits them, taking into account the money spent in their vicinities by amateur hunters."[43] To this end, some regulations termed as poaching any capture that did not embrace the sportsmen's culture and code of conduct. Whether they were hunting and fishing for food, to supplement agroforestry activities, or as the basis of a commercial activity, certain residents of local communities –

108   Surveillance and Improvement of Fish and Game Territories

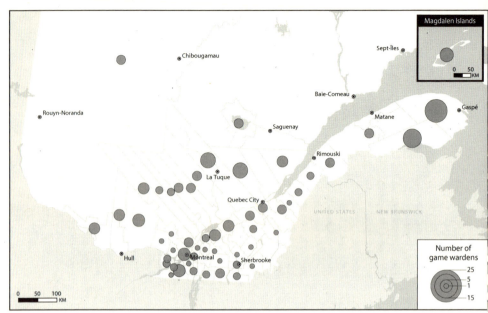

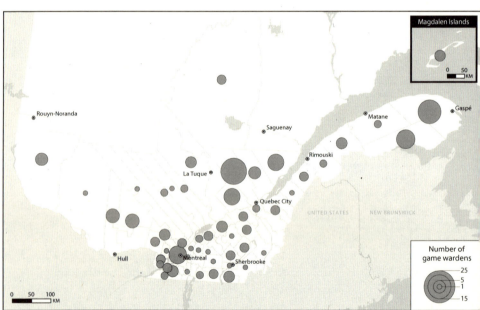

FIGURE 4.6 Division and distribution of Fisheries and Game Service personnel assigned to surveillance by county, 1922 *(top)* and 1932 *(bottom)*. | Sources: *État des comptes publics de la province de Québec*, 1922 and 1932.

subsistence farmers, commercial fishers, and Indigenous people – had to grapple with a new cultural order as the government deemed hunting and fishing exclusively sporting activities and declared all other methods of capture illegal.[44] For example, the 1888 statute, known mainly for the "three chain" rule, forbade all fishing practices not associated with sporting activity in the province's inland waters: "*Angling* (pole and line) ONLY is authorized in the lakes and rivers under the control of the Government of the Province of Quebec."[45] As they were no longer available for trade or subsistence, freshwater fish became a resource to exploit for the pleasure of sportfishers and for development of a tourism industry.

The measures intended to encourage the development of sportfishing and hunting thus led to the criminalization and gradual marginalization of people in Indigenous and rural communities.[46] Although some individuals turned to the courts or submitted petitions to Parliament to demand recognition of their hunting and fishing activities, most simply ignored a legal system that was fundamentally opposed to their methods of capture.[47] Those who fished with banned methods, such as nets and harpoons, became, in the eyes of the law, poachers who were defying the authority of the state. In the view of sportsmen, hunting and fishing with such methods not only broke the law but constituted an immoral act: poachers were seen as individuals who acted selfishly, against their own interest and that of the community, by destroying fish and game and depleting wildlife resources. In addition to denouncing what they considered to be irresponsible behaviour, sportsmen deemed that these methods infringed on the freedom of those who obeyed the regulations and paid fees for the privilege of extracting the generosity of nature, just as trespassing on someone else's property did.[48]

Leasing of Hunting and Fishing Territories: Public Reserves

Historians Darcy Ingram and Darin Kinsey have linked the drop in "poaching" to the abandonment, by some Indigenous and rural populations, of the traditional hunting and fishing practices that the regulations on exploitation of wildlife resources had banned.[49] At the turn of the twentieth century, the availability of inexpensive equipment and the spread of popular magazines on sport hunting and fishing made it easier for people to adopt the sportsmen's code of conduct and change their behaviour. More and more amateurs were forming private clubs and obtaining fish and game estates – certainly smaller in size than those leased to the North American industrial and political elite, but still accessible to this new segment of the population, thanks notably to the expansion of road and railway infrastructure.[50] In addition, the Quebec government was working to put portions of territories into reserves to make them accessible to all sportsmen living in the province without their having to pay the fees ordinarily required to be a member of a club and use its land. To encourage city dwellers to take up sportfishing and hunting, in 1895 the government created Parc de la Montagne-Tremblante, north of Montreal, and Laurentide National Park, north of Quebec City. In fact, these parks were also forest reserves under the Department of Lands and Forests, which was concerned mainly with satisfying the needs of companies holding timber limits within their borders. Furthermore, conservationist objectives with regard to protecting fish and game sometimes conflicted with exploitation of the forest resource, as forest cover was needed to protect wildlife habitat and regulate river flows. This was not the first time that the timber industry and the fish and game industry had faced off, as successive superintendents of fisheries and game had criticized the timber industry's lax stewardship of the rivers: companies dumped prodigious amounts of sawdust into watercourses and neglected to install bypasses to enable salmon and other fish to swim upstream when they were obstructed by dams.[51] Therefore, during his mandate as head superintendent of Laurentide National Park, W.C.J. Hall was constantly torn between the objectives of protecting wildlife and stimulating exploitation of the forest, the latter of which he was overseeing simultaneously in his role as fire prevention superintendent in the Department of Lands and Forests.

Because of difficulties related to broader public access to fish and game territories outside of the private clubs, the superintendent of fisheries and game regularly reiterated the need to create public reserves. In 1901, the department reduced the number and size of tracts to be leased to private clubs in order to establish reserves near colonization regions where a large part of the territory was already leased, such as the Gaspé Peninsula and the north shore of the Ottawa River in Pontiac and Ottawa counties, as well as on the western flank of Laurentide National Park north of Quebec City, along the Quebec and Lake St. John Railway line.[52] Superintendent Joncas had been entertaining this project for two years when he proposed that reserves be created in Témiscamingue, the Lake Kippewa region (in Ottawa and Pontiac counties, almost all the land along the Ottawa River having been leased), the Matapédia Valley, and Témiscouata, to make them into sites for fish and game reproduction. Joncas reasoned that unlike the national parks, which could not be sufficiently protected because they were too big, these smaller reserves would be easier and less expensive to supervise.[53] It was not clear, however, whether they would be open to the public for sporting activities, and thus be regulated, or would serve exclusively for protection of wildlife resources – in which case sportsmen were unlikely to be able to draw any sort of enjoyment from them. Joncas proposed that the establishment of private leases on land adjacent to the reserves be banned and, as needed, that existing leases be cancelled. Finally, it must be noted that Joncas targeted precisely these territories in his 1900 report as being those in which "poaching has become a dangerous plague," a verdict that Caron reiterated in 1907, when he declared those counties to be places where, "every year, we are notified of shameful massacres of our large wild animals perpetrated by poachers and some settlers."[54]

Because the Quebec government was interested primarily in satisfying the demands of business circles to develop the province's resource economy, particularly in the forestry sector, initiatives by the Department of Lands and Forests in terms of public fish and game reserves remained few and far between. Only one was created, in Pontiac County, from former timber limits north of the Ottawa River, in 1902. The department abandoned it five years later.

There were, however, other territories that the Department of Lands and Forests wished to see play a role in conservation of wildlife resources. For instance, the "forest, fish, and game reserves in the Gaspé," established in 1905 in the mountainous part of the peninsula, had a mandate similar to that for the Montagne-Tremblante and Laurentide parks. Although the department claimed to be making available "new and large hunting territories that are not inferior to any other in the Province for the abundance and variety of game," the park in the Gaspé was mainly a reserve to protect forests and farmland in the region against deforestation and erosion. The reserve included the main sources of the salmon rivers in the Gaspé – Grande Cascapédia, Darmouth, York, Saint-Jean, and Bonaventure – which formed, in a way, the original centre of the private club system.[55]

Parc de la Gaspésie provides an illustration of how the Quebec government ultimately saw the establishment of reserves and parks as a means of appeasing the demands of a broader sporting community without having to substantially or significantly modify its administration of the fish and game territories. The park was created in the wake of protests by the growing class of sportsmen – distinct from the class composed of the economic and political elite – who were demanding broader access to wildlife resources. To the argument concerning the costs of this intervention, opponents of private clubs responded that revenues from leasing of territories and sale of licences would pay for surveillance by state wardens.[56] Both members of private clubs and representatives of the Department of Lands and Forests saw this proposal as untenable. During a speech to the convention of the North American Fish and Game Protective Association, Joncas acknowledged that "the leasing of fishing and hunting territories in our province has done more for protection than an army of constables."[57] The granting of these privileges was an effective and affordable way to protect the province's immense territory, which would otherwise have been impossible to monitor properly. Similarly, tenure holders highlighted their investments in building lodges, opening roads, restocking rivers that had been fished out, and hiring wardens to monitor and protect wildlife. Attendees at the convention resolved to adopt a motion regarding the need to make the fish and game territories accessible to the greatest number of people possible by asking the Department of Lands and Forests to offer lots of under ten acres for this purpose.[58]

Laurentide National Park offers another example of the way in which parcels of Crown land were removed from the public domain to form forest reserves and promote the protection of wildlife resources. There was a notable difference, however: in parallel with development of this park, the privileges of private sportsmen's clubs were expanded. In 1905, ten years after the park was created and the same year that the reserve in the Gaspé came into being, the Department of Lands and Forests annexed land "with a view to preserving the forest, the fish, the game, and the watercourses" and allowing access only to holders of park licences, normally granted to employees of companies that held timber limits in the park.[59] Although almost 95 percent of the park's territory was leased to lumbermen, the Department of Lands and Forests renewed the leases granted to the Triton, Tourilli, and Penn clubs because, according to the superintendent of parks, "these clubs guard their privileges jealously and their members observe the hunting and fishing laws to the letter."[60] The department granted five more leases to protect the Grands Jardins sector, situated to the west, where caribou were plentiful. The superintendent regularly recommended the annexation of regions contiguous to the park, which were overrun by poachers, "to place this territory under the control of the statute concerning the Park."[61] He intended to "lease a few pieces of frontage to certain clubs and ... thus complement the protection service for the entire Park."[62] Initially 6,555 square kilometres, the area of the park gradually increased to almost 9,582 square kilometres by 1911.[63]

Given the low numbers of public wardens, how did this extension of the borders of the park contribute to protection of wildlife? In fact, from its creation – in its very design – the park was bordered by hunting territories leased to private clubs (see Figure 4.7).[64] This meant that access to the park was blocked along its northern, western, and southern edges due to the leasing of tracts to clubs that were responsible both for hiring wardens and

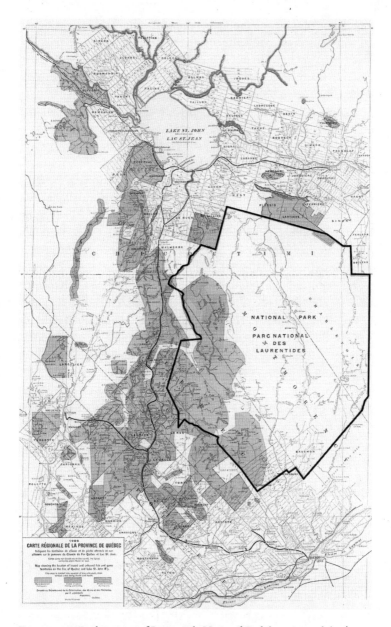

FIGURE 4.7 Enclavement of Laurentide National Park by private clubs. | Source: BANQ-Saguenay, E53, S2, P65, "Carte régionale de la province de Québec indiquant les territoires de chasse et de pêche affermés ou non affermés sur le parcours du chemin de fer Québec et Lac St. Jean = Map showing the location of leased and unleased fish and game territories on the line of Quebec and Lake St. John R'y," P. Jobidon, Québec (Province), Département de la colonisation, des mines et des pêcheries, 1908.

for surveillance of access to the park.[65] In addition, because anyone wishing to enter the forest had to have a park licence, including members of private clubs, there was no need to catch poachers red-handed to arrest and charge them, as individuals were authorized to be in the park and the private reserves only if they had such a licence in their possession. Having the park enclaved by fish and game territories leased to private clubs guaranteed wildlife protection. After the First World War, administration of the park developed in other ways, but these were aimed at protecting the fauna rather than offering the public the opportunity to hunt. In his annual report, the minister constantly repeated his desire to see the province's population become interested in the wildlife within its territory, and he took the opportunity to note the growth in numbers of members of the public taking part in sport hunting and fishing. However, the growing popularity of these activities and the strong presence of American sportsmen exerted pressure on the availability of sites, especially because most rivers were leased and the Fisheries and Game Service did not know where to send tourists who inquired about accessible sites. Therefore, the superintendent of parks encouraged the minister to take advantage of the end of leases on rivers to integrate this land into the public space and "keep it open to all those who enjoy fishing."[66] Since there was no question of challenging the privileges accorded to the minority who held leases, the department worked, instead, to "organize Laurentide National Park in such a way as to form a fish and game reserve and a place for public enjoyment and recreation, in harmony with the spirit of those who created it."[67]

During the 1920s, the Fisheries and Game Service returned to the policy started under Joncas, suspended during the war, of building hunters' camps and lodges and improving travel in the park. Protection of fauna and flora was explicitly added to forestry operations as an objective of the national parks. In 1928, the government banned hunting in the park, making it a sort of game sanctuary – "one of the largest in the country," as the superintendent of parks, J.-Adolphe Bellisle, proudly announced in the introduction to his 1930 report. A few years later, Bellisle vaunted the ecological and economic benefits of this new vocation for the territory:

> Thanks to the protection enjoyed by the fauna there, different animal species flourish in total safety, creating a site that is very attractive to tourists and those who are interested in seeing wild animals in their habitat. In addition, when certain regions of the Park reach their saturation point, so to speak, the game will migrate, which will add value to all the neighbouring wooded regions, filling their hunting territories with game and making them even more attractive to sportsmen.[68]

As the global economic crisis began to hit, the department relaxed its application of the regulations regarding access to the park, illegal exploitation of the forest, and hunting and fishing. The park was prey to poaching, which could also be understood as a subsistence activity for needy populations. Once the hard years of the Depression had passed, the department undertook to "apply the law strictly" and to "gradually eliminate poachers' visiting the park."[69] It also sought to extend the conservationist approach to other areas that it wished to develop into national parks and to the people who might want to enjoy these spaces and their resources. Unlike what took place in Laurentide National Park, the idea was now to make such areas into "a sort of open-air museum where nature must remain unchanging" and where "hunting would never be permitted, in order to allow the public to better observe and study terrestrial fauna, although fishing may be authorized and controlled everywhere that such a thing is possible."[70] In 1937, the government abrogated the 1905 statute establishing the forest, fish, and game reserve in the Gaspé to turn it into a national park. In the view of deputy minister Louis-Arthur Richard, "removing a territory from forestry exploitation – or, at least, strictly regulating cutting – banning hunting, fishing, and mining, and encouraging the public to observe, enjoy, and relax are the principles behind management of national parks."[71] It was not until 1939 that the department developed a hunting site for sportsmen who did not belong to a private club. A ministerial decree dated June 1 of that year created a first public hunting and fishing reserve, the La Vérendrye Reserve, at the border between the Laurentians and Abitibi, in a "region with a large quantity of game and sprinkled with lakes teeming with fish, measuring 2,600 square miles and under the control of the department."[72] As for the reserves created starting in 1905, no provision was made to facilitate access to them. Of the initial objectives that had been assigned to them, only forestry operations continued to be honoured.

The Shaping of the Aquatic Landscape: Pisciculture and Fish Biology

In addition to leasing and surveillance of the fish and game territory, selective interventions with respect to the animals to be hunted and fished were deployed by sportsmen and state agents to establish government of wildlife resources. For example, private clubs encouraged hunters to systematically target certain species – whether predators such as wolves or nuisances such as beavers – even if hunting them was "not a sport" in the view of superintendent Caron and many sportsmen.[73] The case of beavers was particularly interesting, as the creation of the Laurentide National Park reserve was specifically intended to protect them. In his early reports, the superintendent of parks boasted about the growth in the beaver population, but several years later he was noting the complaints made by private clubs that beaver dams were causing floods and destabilizing river flow. The beavers were pointedly blamed for harming hunting and fishing, because they destroyed both fish spawning grounds and sources of food for large game. Caron recommended the adoption of an order-in-council to permit the taking of beavers, even though he recognized the need to apply such permission with caution to avoid a massacre.[74] This measure was not very effective, however; over the years, the clubs often complained about the damage caused by the beavers on their land. Attempts at shaping the territory by manipulating the fauna found another manifestation through the relocation of species within the park and the introduction of species such as the wapiti – although, in the latter case, the superintendent of fisheries and game had to announce one year after the attempt that the animals did not want to remain in the area.[75]

Unlike cynegetic resources, for which attempts at shaping the hunting territory seemed to bear little fruit, ichthyic resources benefited greatly from interventions aimed at improving the aquatic landscape. The state became actively engaged in pisciculture (the cultivation of fish by artificial means). Highly developed in Europe, pisciculture made its appearance in North America in the mid-nineteenth century.[76] The superintendent of fisheries in Lower Canada, Richard Nettle, conducted his own pisciculture experiments with brook trout in a pool created behind his house in Quebec City and then suggested that the colonial administration undertake artificial stocking of the rivers to counter the decline in the salmon population.[77] It was under Nettle's successor, Samuel Wilmot, however, that pisciculture developed on a large scale in Canada. When he was appointed deputy inspector of fisheries at the Dominion Department of Marine and Fisheries in 1868, Wilmot was already conducting pisciculture operations at his home in Newcastle, Ontario, in an attempt to regenerate salmon populations in the tributaries to Lake Ontario.[78] In 1876, the department appointed him superintendent of fish culture and asked him to establish hatcheries across Canada.[79] By the end of Wilmot's mandate in 1895, the department had fifteen hatcheries, several of which were producing sportfishing species such as trout and salmon.[80]

Like the system of private fishing clubs and surveillance of the wildlife territory, fish production became a component of the regime implemented under the administration of the Province of Canada that was adopted by the Quebec state. In August 1915, the federal Department of Marine and Fisheries ceded its hatcheries to the provinces on condition that they be used for the conservation of salmon populations and that they distribute fry free of charge.[81] The transfer of hatcheries to provincial control took place at the same time as the transfer of rights on management of inland waters. Quebec obtained responsibility for the hatcheries situated at Mont-Tremblant in the Laurentians, Saint-Alexis-des-Monts north of Trois-Rivières, and Baldwin Mills and Magog in the Eastern Townships; in 1915, these hatcheries released more than two million trout and salmon fry into the province's inland waters. Then, in 1922, when the federal department ceded "control and administration of all fisheries in the province's waters" to the Quebec state, the salmon hatcheries in Gaspé and Tadoussac joined the provincial Fisheries and Game Service. They were added to four other pisciculture facilities, which raised mainly speckled and rainbow trout.

The federal hatcheries, which Caron initially saw as a financial burden to his service, quickly became a driving force for the strategies to stimulate sportfishing and the tourism industry.[82] Caron appointed E.T.D. Chambers manager of the provincial hatcheries in 1916. Since 1911, Chambers had been acting as a special officer of the department charged with investigating provincial hunting and fishing laws.[83] Within the North American sportsmen's community, he was well known for having self-published many books vaunting Quebec's high-quality hunting and fishing territories.[84] The hiring of Chambers, whose writings on sportfishing were popular throughout the continent, signalled the department's desire to connect fishing to the tourism industry. He undertook to sell fry from the provincial fish farms to private clubs rather than distribute them free of charge, contravening the commitment made by the province to the federal administration when the hatcheries were transferred.[85] He also introduced species prized by the members of private clubs who came to Quebec from all over the world to appreciate the wealth of the sportsmen's paradise: rainbow trout, Atlantic salmon, and brown trout fry were spawned in the facilities of the Hatchery Service to stock Quebec's fishing rivers. The service became financially self-sufficient in just two years, as demand from private clubs and individuals grew beyond the capacity of the hatcheries. In 1917, 4 million fry and fingerlings were sold.[86]

The Hatchery Service had been breeding fry for more than a decade in April 1930, when the Department of Colonization, Mines, and Fisheries hired Bertram William Taylor, a biologist with a master's degree from McGill University. One of Taylor's first tasks was to assemble a biology service to perform a hydrobiological and ichthyological inventory of the province's watercourses that would form the basis of the activities of the Hatchery Service.[87] Two years after he was hired, Taylor succeeded Chambers as head of the Hatchery Service, which he ran along with the Biological Service. Taylor extended the network of provincial hatcheries by establishing stations at Lake Manitou in the Laurentians, where he started up a program for inspection of lakes and analysis of their chemical and biological conditions, and at Saint-Félicien, near Lake St. John, for production of freshwater salmon (known as "ouananiche") to supply the entire network, which otherwise produced essentially salmon and trout fry.[88] Taylor redefined the operations of the Hatchery Service, which up to then had unquestioningly followed the orientation that the federal Department of Fisheries had impressed upon provincial hatcheries: production of trout and salmon with which clubs could stock their tenured lakes and rivers.[89] After early studies were conducted at Lake Manitou and Brome Lake, near the hatcheries at Mont-Tremblant and Baldwin Mills, respectively, Taylor asked that production of Atlantic salmon fry for stocking freshwater bodies be stopped, given results that he found disappointing. He was also skeptical about the survival capacity of rainbow trout raised in the hatcheries at Magog and Baldwin Mills in the Eastern Townships. His evaluations contradicted those that had been issued by his predecessor several years earlier. Chambers had asserted: "The different rivers stocked with fish by us have been inspected, and their tenure holders or guardians agree that the stocking that we are doing is to the advantage of the rivers and of salmon fishing in general, and that a large percentage of the increase in fishing in recent years can be attributed to this fish stocking."[90]

Taylor envisaged, instead, promoting indigenous species at the expense of the two salmonid species that had been pleasing sportfishers for decades and to which the Hatchery Service had devoted most of its resources. He

FIGURE 4.8   B.W. Taylor stocking Lake Sorcier with ouananiche fry, September 27, 1933. | *Source:* Author's collection.

had certain facilities specialize in the raising of small-mouth bass and ouananiche, indigenous species for which the environmental conditions in Quebec lakes would be more appropriate. In the case of the latter species, it must be noted that Chambers had promoted it throughout his career, notably with the publication in the United States of his book *The Ouananiche and Its Canadian Environment* in 1896.[91] As director of pisciculture in the Fisheries and Game Service and advocate of sportfishing and tourism in the province, he had also encouraged the transplantation of ouananiche as a sportfishing species without committing the service to its reproduction and breeding. Taylor, with his scientific background, had other ideas about the Fisheries and Game Service's territorial development activities that included large-scale changes to the ichthyological composition of the lakes and rivers of Quebec. He initiated a program to study lakes in the Lac-Saint-Jean district in 1930, but, without waiting for the results to come in for this "territory that is completely unknown to us from this point of view [report on the waters]," the Hatchery Service began to raise ouananiche at Saint-Félicien and use them to stock Lake St. John and its tributaries and the service's network of pisciculture stations (Figures 4.8 and 4.9).[92] Similarly, at the Magog station, where, according to Taylor, "the water is not appropriate for hatching [rainbow trout] eggs and raising them to the state of fingerlings and fry," the construction of tanks for raising small-mouth bass fry began the following year to increase capacity to breed this indigenous fish species.[93]

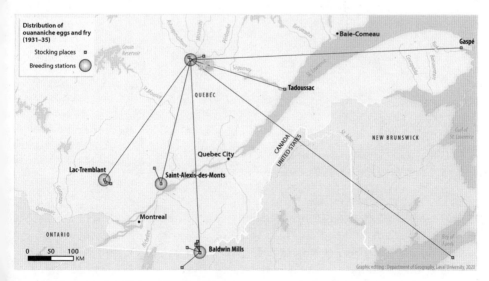

FIGURE 4.9 Distribution of ouananiche eggs and fry in Quebec, 1931–35. | *Source: Rapport annuel du service de pisciculture, 1932–35.*

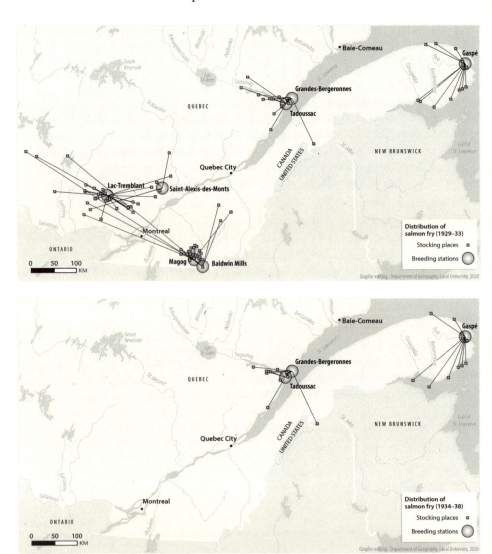

FIGURE 4.10 Distribution of salmon fry in Quebec, 1929–33 *(top)* and 1934–38 *(bottom)*. | Sources: *Rapport annuel du service de pisciculture*, 1926–30, 1932–34, 1936–37.

Diversification of fish production in the provincial hatcheries lasted a few years, but in the second half of the 1930s, Taylor turned the activities of the Hatchery Service toward regional and species-specific specializations. The service stopped raising lake trout entirely and confined rainbow trout to a single station; it also abandoned bass at the Baldwin Mills station and ouananiche at the Saint-Félicien station, which ceased operation after 1935.

Instead, it opted to specialize its stations mainly around farming salmon and speckled trout. In addition, as the service's hatcheries were engaged in identical production, their scope of action decreased dramatically: their stocking activities were limited to nearby counties rather than targeting the supply of different regions (see Figure 4.10). For example, the stations in the Eastern Townships (Magog and Baldwin Mills) supplied watercourses situated between the Richelieu and Chaudière Rivers and stopped sending salmon fry north of the St. Lawrence River. Stocking the watercourses between the Ottawa and the Saguenay Rivers became the responsibility of the Saint-Alexis-des-Monts and Saint-Faustin stations, to which the Hatchery Service transferred the activities previously performed at the Mont-Tremblant and Lake Manitou stations (see Figure 4.11). In addition, the service limited the principal production of these two stations to speckled trout.

FIGURE 4.11 After small experiments with raising speckled trout in 1933, the Saint-Faustin pisciculture station gradually absorbed the activities of the Mont-Tremblant station in 1934 and the Lake Manitou station in 1936. It expanded under Gustave Prévost's leadership and was provided with scientific equipment when Prévost became the first director of the Biology Bureau in 1943. | The provincial fish hatchery at Saint-Faustin in the Laurentians: an aerial view of pisciculture, by Lucien Piché, 1944. *Source:* BANQ-Québec, Fonds ministère de la Culture et des Communications, Office du film du Québec, Documents iconographiques, E6, S7, SS1, D18187.

Another orientation that Taylor impressed upon his organization, which was maintained during the 1930s, involved a close relationship with academic researchers and the regular publication of scientific and technical articles by the services' personnel in the most important North American journal on the topic, *Transactions of the American Fisheries Society*. For ichthyological reports on Quebec lakes, Taylor drew on his alma mater's network, including bacteriology professor R.E. Jamieson and zoology professor V.C. Wynne-Edwards, who had recently arrived at McGill University, to take charge of similar work in the Eastern Townships.[94] It was, however, with the Institut de zoologie at the Université de Montréal that the Biological Service developed the closest research ties. In 1938, the department subsidized construction of the biology station at Lake Jacques-Cartier, in Laurentide National Park, to map the hydrography of the park's lakes and conduct a study on brook trout.[95] Appointed director of the station was ichthyologist Vadim D. Vladykov, a professor at the Université de Montréal, who led a team of students and researchers in various biological disciplines there. Gustave Prévost, a researcher attached to the pisciculture station in Saint-Faustin, also taught in the biology department at the Université de Montréal. Prévost collaborated with another professor from the university, Lucien Piché, to improve transportation and release conditions for fingerlings to stock watercourses.[96] In addition to making it possible to assess the quality of fish-stocking work by studying the province's lakes and rivers, the expertise developed within and around the Biological Service provided the Quebec state with tools for apprehending phenomena affecting the state of wildlife populations, the degradation of aquatic environments, and the depletion of fish resources.

THE QUEBEC FISH AND GAME territory, which took shape under the administration of the Province of Canada, involved decisions affecting the granting of fishing territories in the Gaspé, the delegation of improvement and surveillance work to private sportsmen's associations, and large-scale fish production. Legal and constitutional developments that determined the separation of powers and the rights to Crown lands led the Quebec state to extend this approach to all its watercourses, then to all its hunting territory. Through regulation of private clubs' lands and creation of public reserves and infrastructure, a code of sporting conduct formulated by members of the industrial and economic elite was inscribed in the hinterland landscape for users of the fauna from different backgrounds and classes, contributing to the establishment of province-wide government of wildlife resources.

Despite the cost of monitoring its wildlife estate and running the Hatchery Service, the Quebec state's stewardship of its sportsmen's paradise provided the public treasury with substantial revenues through leases on hunting and fishing territories and the sale of licences. However, the pisciculture and fish-stocking activities led not only to changes to the wildlife landscape but to reconsideration of the terms of action for the Fisheries and Game Service. Rather than simply setting out to monitor the territory and its users or to intensify production at its hatcheries, the service sought to enhance its improvements through fieldwork and mobilization of academic researchers. In doing this, it created an institutional niche for development of wildlife biology research, which was the basis for modernization of its intervention mechanisms.

5
Regionalization and Specialization of Agricultural Production: Disseminating Agronomic Knowledge

During the nineteenth century, the reorientation of agricultural production became a major concern for commentators who wanted to see Lower Canadian agriculture recover from its "crisis."[1] Declining wheat yields bound for export, blamed on low availability of cultivable land due to population growth or the use of routine practices that exhausted soil fertility, led a number of these commentators to opine on the future of farming. Publication after publication – starting with a report by a commission on the state of agriculture chaired by Joseph-Charles Taché in 1850 – recommended the adoption of "proper techniques" and "suitable crops."[2] The notion of suitable crops was particularly interesting, as one way to make them suitable would be to grow them under propitious agro-ecological and commercial conditions. Wheat crops were becoming less and less profitable, as producers in the western part of the country held the decisive competitive advantage of richly fertile soils newly put into cultivation. Commentators therefore recommended that farmers turn to crops that would adapt well to local soils, climates, and markets. Because it led to the shaping of the agrarian landscape and the formation of specialized agricultural producers, as well their integration into a market economy, the introduction of suitable crops provided leverage for the government of agricultural resources.

From the hiring of agricultural speakers in the second half of the nineteenth century to the appointment of official agronomists starting in 1913, the Quebec Department of Agriculture encouraged this reorientation by financing educational activities and by disseminating agronomic knowledge and techniques. Nevertheless, the department's interventions changed profoundly during the period studied, as it moved from basing its educational enterprise solely on the associative movement to building a corps of professionals to maintain constant contact with farmers throughout the province. In parallel with assigning responsibility for disseminating knowledge to its own agronomists, the department adopted a more active role in the framing of agriculture to encourage the growth of dairy production and the spread of mixed farming.[3] It then helped the formation of specialized producers, which entailed a redrawing of the province's agricultural territory. Indeed, the department had to devise composite interventions to reach out to the various clienteles within the agrarian landscape of Quebec: on the one hand were farmers close to urban centres and transportation routes, engaged in commercial agriculture; on the other hand were farmers, ordinarily situated in new colonization zones on the edge of the ecumene, who persisted with subsistence agriculture.

In this chapter, I examine how the specialization of two types of actors – agronomists and farmers – guided changes to the agricultural map of Quebec that concurrently benefited from the regionalization of production structured by the interventions of the state agents. Provided with direct access to farmers throughout the territory following the creation of an agronomy service, the Department of Agriculture shaped specialized agricultural regions by introducing crops and livestock through its agents or by consolidating production in territories whose contours it expanded to raise both its own visibility and that of the agriculture industry.

Scales of Dissemination of Agronomic Knowledge

After Confederation, the provincial administration based its agronomic action on two institutions inherited from the Province of Canada: county agricultural societies and the Agricultural Council, which replaced the Lower Canada Chamber of Agriculture created in the wake of the Taché report.[4] Composed of members of the urban and rural elite, these two institutions operated hand in hand, as the Agricultural Council provided grants to the agricultural societies to disseminate the latest agronomic advances to farmers. The agricultural societies purchased breeding animals, distributed grain seeds, and organized fairs and contests to encourage emulation of best practices. Agricultural reformers questioned whether these societies were effective vectors of change, however, as they were organized mainly around the most prominent farmers and regular recipients of prizes and awards handed out at contests and fairs.[5] They seemed little, if at all, interested in the less-well-off farmers who were supposed to be the primary beneficiaries of extension and demonstration activities. They also had to cover large territories; even though in some cases two societies shared a single county, few of them managed, even if they tried, to effectively reach the largest number possible; each society's membership was generally limited to some forty farmers. None was able to recruit members in the parishes farthest from its centre of operation or to exert its influence across its entire county. Finally, according to historians Jean Hamelin and Yves Roby, the agricultural societies were prey to conflicts that reproduced oppositions based on membership in a political family or a linguistic group, which did not encourage the participation of outsiders.[6]

Criticisms flew thick and fast regarding the relevance and effectiveness of this form of extension activity. As calls continued for a reorientation of agriculture in the province, some suggested a vehicle other than traditional agricultural societies for encouraging farmers to adopt the new agronomic principles. One possibility was the farmers' club, a recently created parish institution; the first was founded in 1862 on Île d'Orléans, and clubs sprang up around the province to number about forty by the early 1880s.[7] Unlike the agricultural societies, farmers' clubs mobilized their members on a regular basis and in larger numbers, even though they were hosted by the biggest farmers in each parish. In the view of the director of the Department of Agriculture and editor of the *Journal d'agriculture illustré*, Édouard-André Barnard, farmers' clubs were a more appropriate way to reach farmers because they were formed at the parish level, whereas the agricultural societies operated at the county level.[8] It was not simply a question of scale; Barnard used the farmers' clubs to break the monopoly held by members of the agricultural elite who sat on the Agricultural Council and in county societies and dictated the department's operations and policies. He was not shy to criticize these institutions: from 1872 to 1876, when he was the department's inspector of agricultural societies, he denounced the societies' poor bookkeeping and the indifference of the Agricultural Council, which was supposed to oversee the proper use of government grants.

With the support of local religious authorities, Barnard sought to have farmers' clubs established in a large number of parishes, as they were closer to individual farmers than were the county societies. He made sure that farmers joined the clubs and took part in their activities, especially when they hosted the speakers that the department dispatched throughout the province. These speakers were not necessarily agronomists, but all, including members of liberal professions, had studied enough of the subject to be able to disseminate information.[9] They buttressed the clubs' activities by telling farmers about the Department of Agriculture's programs and measures. Barnard also counted on having the speakers' lectures read in the *Journal d'agriculture illustré,* which he published through the good auspices of the department and to which, for a special rate, the clubs were required to subscribe all of their members – even though illiteracy was the rule in rural areas.

In 1893, the commissioner of agriculture, Louis Beaubien, officially recognized the farmers' clubs. In exchange for subsidies, the clubs had to ensure that their activities and administrative operations complied with the department's requirements.[10] Some felt that receiving subsidies meant the end of the farmers' clubs' independence; for example, they had to have at least twenty-five members to receive their annual subsidy. Nevertheless, official recognition of the clubs accelerated their spread through the province: from 530 in 1900, the number grew to 736 in 1918, with a total of 71,173 members.[11]

In addition to freeing the Department of Agriculture from the domination of the elites, this parish institution had a marked effect on the reorientation of agriculture in Quebec, including the adoption of mixed polyculture and the establishment of cheese and butter factories. Moreover, the Société d'industrie laitière de la province de Québec, which Barnard helped to found in 1882 and on whose board he sat, provided unequivocal support to the clubs in 1884 by proclaiming their usefulness to the dairy industry.[12] The Société d'industrie laitière, which managed the École de l'industrie laitière de Saint-Hyacinthe and trained inspectors and dairy-processing factory operators, was a private institution, but it was generously subsidized for operation of the dairy school and inspection of factories. Its creation was one in a series of measures taken by the department, which had decided in 1875 to make the dairy industry the keystone to modernization of Quebec agriculture.[13] Cheese and butter factories sprang up in numerous parishes in the province as farmers' clubs were founded; from 162 in 1881, the number of factories grew to 730 in 1891 and almost 2,000 in 1901.[14] Eventually, there were so many that the quality of the dairy products marketed locally and internationally suffered, but they nevertheless enabled farmers to be involved in a highly profitable activity, either by specializing, as was done in the Richelieu Valley, or by complementing this production with another commercial specialty, such as livestock raising in the Eastern Townships, growing hay in Champlain and Portneuf counties, and market gardening near urban centres.[15]

The clubs and societies were distinguished by the implementation of two scales of agronomic extension activities: the parish and the county, respectively. By financing both associations, the Department of Agriculture showed its desire to work on two fronts to encourage modernization of agriculture. At the same time that it was seeking to consolidate the efforts of large producers engaged in commercial agriculture, the department also hoped that smaller farmers, many of them practising subsistence farming, would improve their production techniques and expand into commercial production, which would provide more than simply a supplementary income. Within colonization parishes, farmers' clubs could also be a more effective tool than agricultural societies for integrating farmers on the settlement frontier into trade circuits. In this sense, the department continued to assume the responsibilities involving colonization that had been conferred upon it when it was created just after Confederation.

Although reformers such as Barnard emphasized the antagonisms between the two types of associations in terms of mission and operations, others underlined their complementarity. They saw the proliferation of farmers' clubs as a prelude to the creation of effectively functioning agricultural societies, as the clubs, which made it easy for small groups of farmers to meet, might eventually federate at the county level.[16] Recommendations that they be merged appeared in 1898, but this came to pass only in the 1940s. In the meantime, the department hired staff to reach out to the agricultural class and continue the recovery of the rural economy.

In 1913, a new "agricultural progress" agent, the official agronomist, came on the scene. At this point, the department stopped leaving educational activities entirely to the various agricultural associations and took charge of regulating the agricultural sector.[17] It started to consolidate the industry by internalizing the skills that the associations, notably those involved with crops and livestock, deployed to serve their members. With financial resources provided by the federal government's Agricultural Instruction Act, it recruited agronomists from the first cohorts of graduates with a bachelor's degree in agricultural sciences at Université Laval and its Montreal affiliates (the École d'agriculture de Sainte-Anne-de-la-Pocatière and the Institut agricole d'Oka) and at McGill University (Macdonald College), and it established a local mechanism for disseminating agronomic information.

Acknowledging that many farmers were relatively unenthusiastic about the activities of the agricultural associations, the minister at the time, Joseph-Édouard Caron, justified the hiring of this scientific personnel as follows:

> There are still too many farmers who rarely go to fairs, demonstrations, or lectures. We have decided to go to them, on their farms, so that none of them can have the slightest excuse not to make progress and adopt better cultivation techniques. We have therefore appointed five agronomists – all graduates of the Institut d'Oka – who will live permanently in the districts assigned to them. They will advise farmers, give lectures, and monitor agricultural associations such as cooperatives, farmers' clubs, and agricultural societies.[18]

The Agronomic Service once again positioned the county as the territorial unit for the dissemination of knowledge. Yet, through its interventions it sought to reach out to both large producers and farmers situated on the settlement frontier, thanks to the mobility of the official agronomist and his permanent residency in the county. Initially, there were too few agronomists to be assigned individually to specific counties, but in 1916, the minister announced that he wanted to set up an agronomy office in every county.[19] In the meantime, the department immediately assigned the first agronomists to five districts, each composed of two counties: J.-S. Simard to Montmorency and Quebec City, Henri Cloutier to Iberville and Rouville, R.-A. Rousseau to Bagot and Drummond, Jean-Charles Magnan to Champlain and Portneuf, and Abel Raymond to Bellechasse and Dorchester. These districts were highly productive agricultural zones, where the conversion to dairy production had been achieved rapidly. It was similar for Cookshire, Huntingdon, Lennoxville, Cowansville, and Richmond, where Macdonald College had already put extension agents in place by hiring its first agronomy graduates using subsidies obtained under the Agricultural Instruction Act.[20] A sixth agronomist, J.-M. Leclair, also a graduate of Macdonald College, was added to the agronomy corps the following year to serve the "beautiful agricultural district" of Abitibi, now open to colonization after the inauguration of the National Transcontinental Railway line.[21]

During the Agronomic Service's first decade, the number of agronomists – and, in parallel, the number of agronomic districts – grew exponentially (see Figure 5.1). To support this growth, the Department of Agriculture used federal grants received under the Agricultural Instruction Act, as well as funding from campaigns during the First World War, to intensify agricultural production to meet the needs of the Allied forces and local populations. In the year 1917–18 alone, fifteen new agronomic districts were added to the total of nineteen that had existed the preceding year. The department thus made twenty-nine of its thirty-four agronomists responsible for a single county, even though twenty-nine counties still did not have an agronomy office in 1920.[22] After the federal government's decision to abrogate the Agricultural Instruction Act in 1923, the pace of recruitment slowed and the department hired one to three new agronomists per year, except in 1929. It must also be noted that as of 1923, the minister could congratulate himself on having reached his objective of one agronomy office per county, with sixty-nine agronomists and several assistant agronomists in position.[23] Far from slowing the activities of the Department of Agriculture, the war had acted as a stimulant. Its effect was to structure the activities of the Agronomic Service in forty counties, some of which counted more than one agronomist.

FIGURE 5.1   Number of official agronomists, 1913–32

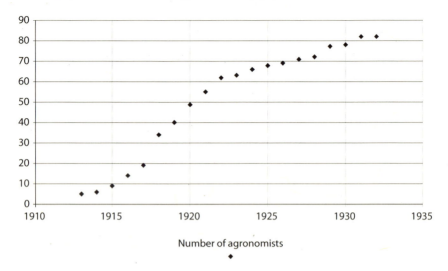

Source: *Rapport annuel du ministère de l'Agriculture (RAMA)*, respective years.

A closer look at the geographic distribution of agronomists highlights certain aspects of the department's educational enterprise. The county, a political division, was "a decisive factor in the administrative structures" of the Department of Agriculture, and thus of its agronomic activity, but other considerations, such as the spatial disparities of agricultural production, had to be taken into account.[24] For purposes of comparison, I will focus on three years: 1917, before the sudden rise in number of agronomists; 1923, when the Department of Agriculture had agents in all counties and, in any case, funds were lacking after abrogation of the Agricultural Instruction Act; and 1932, the year before the agricultural map of Quebec was redrawn into twenty agronomic districts. After that, figures were no longer compiled systematically in the department's annual report and, at the same time, "technical specialists" began to be designated as official agronomists.

In 1918, the department presented a "map of the province of Quebec showing the agronomists' districts" (Figure 5.2) in its annual report. With the exception of Abitibi, where the department had considered it a good idea to install an agronomist as of the second year of the service, and Champlain and Portneuf, where the farmers in each county were served by an agronomist, the counties east of the Richelieu River, on the south shore of the St. Lawrence, and east of the Maskinongé River on the north shore, were paired to share

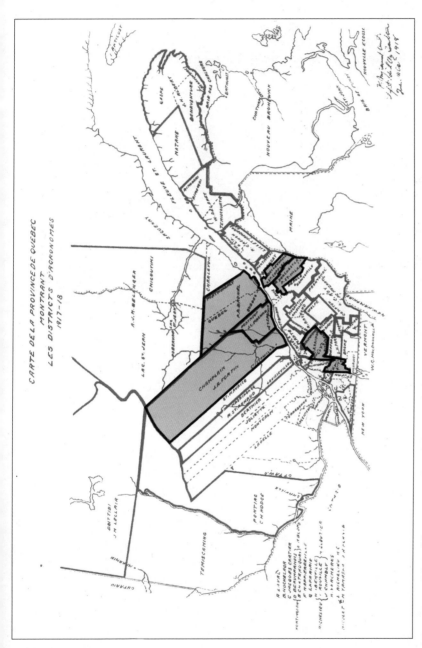

FIGURE 5.2 Map of the province of Quebec showing agronomic districts, 1917–18. The first agronomic districts, formed in 1913, are shaded. | *Source: RAMA*, 1917–18, 2.

an agronomist. To the west, two agronomists were responsible for six counties south of the island of Montreal. At the time, the farms in Jacques-Cartier and Laval counties were among the most productive in the province, along with those in the Richelieu Valley.[25]

Although the early counties selected by the department to welcome an official agronomist seemed to reinforce the idea that public assistance generally benefited the most prosperous farmers, the Quebec governmental policy thereafter would call for a more nuanced interpretation. Indeed, the department turned its efforts toward marginal zones with less productive soil, independent of what was grown or raised or the techniques used. In 1923, the addition to the electoral map of new counties in Abitibi, the Matapédia Valley, and the upper Laurentians north of Montreal was accompanied by the hiring of agronomists to be assigned to them. The department also added agronomists in Lac-Saint-Jean and Champlain, where the backcountry was being settled. When these counties were subdivided a few years later, agronomists were already in position to serve farmers in the new counties of Roberval and Laviolette. Throughout the province, each county had its own agronomist, although the more prosperous agricultural territories, also smaller in size than the average county – such as Terrebonne and Laval, Napierville and Laprairie (which together formed a single county), Vaudreuil and Soulanges, Richelieu and Verchères, and Saint-Jean and Iberville – shared an agronomist.

In 1932, the geographic distribution of agronomists indicated that agronomic activity continued to follow the political divisions of the territory, with departmental agents being hired as new counties were created; this was the case for Gatineau, Laviolette, Roberval, and Rivière-du-Loup. Only in the Montreal plain did some counties (Rouville-Chambly, Jacques-Cartier–Laval, Saint-Jean–Iberville) continue to share an agronomist, although others (Vaudreuil, Soulanges, Verchères, Richelieu, and Terrebonne) now had their own agronomist. Finally, some counties in central Quebec on either side of the St. Lawrence, in the Gaspé, in the upper Laurentians, and in the northern part of the province (Abitibi, Roberval, Charlevoix) had two agronomists. The department invoked the size of these counties as a reason to divide them and assign an agronomist to each subdivision, although the colonization projects launched during the Depression also played a role in this decision.[26]

Regionalization of Agronomic Activity

Even as it brought political advantages, given the heightened visibility for the government afforded by the constant presence of Department of Agriculture agents, the support provided to farmers in colonization territories seemed at odds with the political discourse regarding growth and consolidation of a national industry to be built on technical supervision of large producers. Other considerations must be taken into account in analyzing the spatial distribution of official agronomists. Up to 1897, the Department of Agriculture offered assistance to farmers in remote colonization areas, first by being responsible for public works (1867–86) – thus for the construction of roads and colonization routes – and then by taking over the colonization sector (1887–96). Moreover, by staying in touch with these farmers and seeking to integrate them further into trade circuits and the market economy, the department was not neglecting commercial agriculture – far from it. Below, I describe how the department's services shaped the growth of the fruit-growing and poultry sectors, notably by internalizing the skills developed by associative actors in these sectors. In addition, an examination of the organization and orientation of the agronomists' activities shows that the department supported regionalization and consolidation of the agriculture industry. Among other things, the distribution of agronomists in 1932 reveals their presence in a number of counties previously neglected by the department, notably in the Montreal plain, the "real heart of the Quebec agrarian space."[27] Other department initiatives regionalized the agronomists' interventions in order to encourage specific types of production in the rural landscape.

A first attempt at regionalization of agronomic activity was made once the objective of "one county, one agronomist" was achieved. Caron then redrew the agricultural map of Quebec by grouping the agronomists in districts.[28] The reorganization was aimed at consolidating the work of inspectors responsible for supervising and coordinating the agronomists in regions with set contours but containing a fluctuating number of counties. In December 1922, the director of the Agronomic Service, François Narcisse Savoie, announced the formation of a first agronomy district "composed of a specific number of counties"; the district covered the Montreal region and comprised twelve county agronomy offices.[29] The department envisaged establishing five more agronomy districts province-wide, but it is difficult to know exactly how it intended to define them, because their borders underwent numerous revisions as counties were added or subtracted. In 1926, two more districts were created: Bas de Québec (no. 1) and Cantons de l'Est (no. 5); two years later, Trois-Rivières (no. 4) came into being. On October 1, 1929, the department stated that all districts were operational, but there were only five of them.[30] In the spring of that year, Premier Taschereau had replaced Caron as minister of agriculture with Joseph-Léonide Perron, who undertook to redraw the agronomic map of Quebec on new bases.[31]

Perron intended to establish "an energetic policy of agricultural renewal" by reorganizing the department around agricultural regions defined according to their agro-ecological and economic characteristics.[32] Displeased that so many imported foods were to be found on family tables and in market stalls in Quebec, Perron wanted to improve production and marketing of agricultural products in the province.[33] Under a program formulated by three service heads but first articulated in the newsletter of the official agronomists, *Le Lien,* in 1928, agronomists became responsible for investigating "the determination of different types of soils in each county and their location in relation to the parishes," as well as the "main crops grown and related production on farms."[34] The program included preparation of an agricultural map of the province drawn from statistical information concerning the agricultural realities in counties and "making it possible to group together counties or parts of them, taking account of soil conditions, climate, markets, and transportation, that will cause cultivation systems to vary in such a way as to group farmers for the purposes of production and sales."[35] The map resulting from this exercise divided "the area of arable land currently under cultivation" into twenty-three agricultural districts (Figure 5.3). Perron wanted to "be able to provide effective guidance and tell farmers what they have to produce, informing them about the markets within their reach."[36] In the agronomic districts, it was the agronomists who were entrusted with this mission, and their "main work will consist of directing cultivation, taking account of the main needs of each region."[37] Perron's hope was that "these regions where cultivation systems are similar will create production centres to supply businesses."[38]

Although the department seemed to be naturalizing agricultural regions by designating specialized types of production as a function of "regional characteristics" to complement the existing mixed polyculture system, it recognized that the productivity of certain yields depended on environmental transformations such as soil improvement through drainage and the use of manure and lime. Moreover, this "naturalness" was based on an extensive integration of regional agriculture with urban markets, organization of sales and marketing of agricultural products, and specialization of regional production as a function of agronomic characteristics. With the motto "Let us produce and sell properly, according to regional specificities," the policy was, in part, a response to an agricultural crisis that was seeing tariff barriers multiplying in foreign markets; it encouraged "buying local," "conquering markets," and improving, increasing, and specializing agricultural production.[39]

The 1929 agricultural map was the only one on which the territories of agronomic activity were detached from the political lines of Quebec's electoral

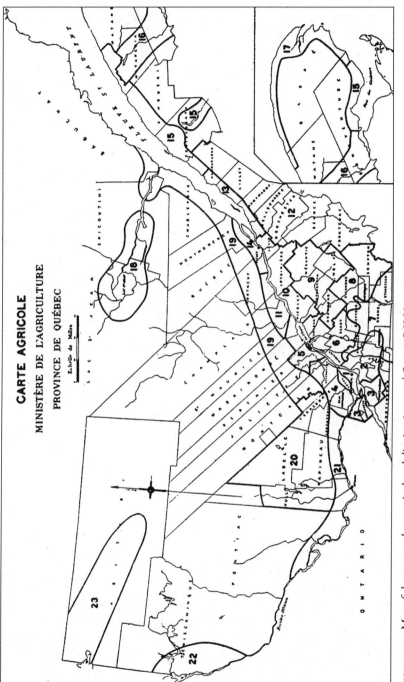

FIGURE 5.3 Map of the twenty-three agricultural districts in 1929. | *Source: RAMA,* 1928–29, vi.

map. Although the borders of counties were occasionally aligned with those of agronomic districts, other factors were also taken into account. Notably, the foothills of the Laurentians demarcated the northern limit of district 19 inland. It is also interesting to observe that there were some enclaves (for example, districts 18, 22, and 23 in the northern part of the province and district 15 in the region around the Témiscouata River) and that certain districts were divided among marginal zones inland and more fertile zones bordering the St. Lawrence. These distinctions testified to the existence of different kinds of agricultural production with particular needs, or a "two-speed" agriculture. Among other things, the map increased the number of regions in the St. Lawrence Lowlands in which commercial farms specializing in market gardening and fruit growing prospered thanks to their proximity to urban centres.

During the 1930s, the renewal program continued, although Perron's "agricultural map" was not implemented. Perron died suddenly in November 1930, and he was immediately replaced by Joseph-Adélard Godbout, an agronomist who had graduated from the École d'agriculture de Sainte-Anne-de-la-Pocatière, where he taught animal husbandry.[40] In a context of economic crisis and budgetary restrictions, Godbout rationalized the department's activities and reorganized the province's agricultural territory by dividing the province into twenty agronomic districts based on the political divisions of the counties (see Figure 5.4); a regional agronomist was put at the head of each district "to supervise the work of the county agronomists," a task that had fallen to inspectors after Caron's 1922 reorganization.[41]

Beyond simply serving administrative rationalization – which, paradoxically, led to an increase in the number of districts from five to twenty – the 1933 map helped to bring a national industry based on dairy production to the forefront, while relegating small producers to the background.[42] Some districts corresponded not to the description accompanying the map but to the grouping presented in the 1935 *Annuaire statistique de la province de Québec*, whereas the section devoted to dairy production presented twenty districts, the contours of which corresponded exactly to those presented in 1933.[43] These twenty districts that made up Godbout's map reproduced regional divisions based on statistical studies that the Department of Agriculture had started to conduct again in 1929, when Perron created an agricultural statistics section within the Rural Economy Service (this activity had been abandoned to the federal statistics bureau in 1924).[44] The section gathered data supplied by farmers on "formula cards" that the department distributed to them through its official agronomists: in 1930, 23 percent of farmers took part in this annual census; by 1932, over 40 percent were participating.[45] The data collected included variables such as number of dairy cows and other

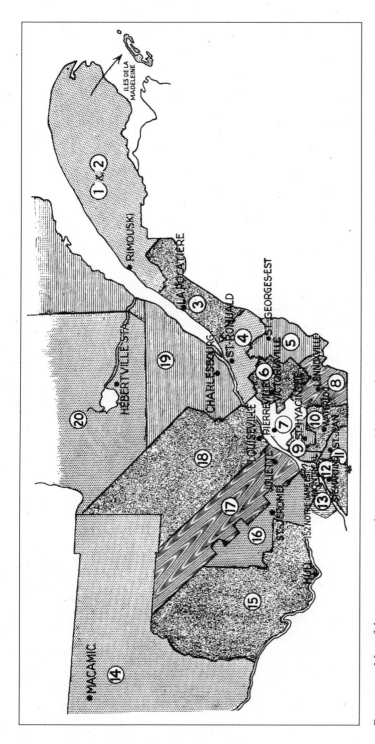

FIGURE 5.4 Map of the twenty agronomic districts in 1933 and location of residence of the regional agronomist for each district.
Source: *RAMA*, 1933–34, 2.

domestic animals, areas cultivated, average crop yields, and selling prices. Gathering this information involved cooperation by the farmers, who were engaging in the bookkeeping activities that the agronomists had been promoting since the late nineteenth century. Because only farms larger than five acres were counted, the department was delimiting the viability and efficiency standards of farming operations run by full-time farmers seriously engaged in a commercial activity.[46] Although some farmers took on other economic activities, it was above all the dairy sector that defined the average farmer around whom the department articulated its technical and regulatory interventions.

Using the divisions on the 1933 map, the department could focus on its agronomists' interventions and gather statistical data on identical territorial bases. The regions that benefited the most from the regional concentration of scientific personnel established in the wake of the cartographic and statistical reorganization of 1933 were those to which the department had sent fewer employees in the past. They were also the ones where dairy farming had been solidly established since the late nineteenth century, and the addition of livestock and crop specialization had made it possible to profit from the proximity of urban markets and sites for export, notably in the region around Montreal and the Eastern Townships.[47] The territories that had received more sustained technical assistance, in which a concern with hinterland colonization had led the Department of Agriculture to expand the presence of its agronomists, seemed to see a drop in numbers of technical personnel after the 1933 reform. The rationalization effort was aimed less at the department and its administration than at agricultural production and commercially successful farmers.

Finally, the 1933 map shows how little Perron's 1929 map seems to have influenced agronomic activity, notably in terms of the collection of statistics from which he had made the map.[48] Among other things, the 1933 map left out small production and self-sustaining farms, whereas the 1929 map had openly displayed two-speed agriculture, which caused discontent within agricultural circles: in effect, it sanctioned the marginal nature of some production and some regions, in the foothills of the Laurentians from north of Montreal to Trois-Rivières and Quebec City, and south of the St. Lawrence, on the American border. Furthermore, by combining counties and agronomic districts, the 1933 map had the advantage of raising the Department of Agriculture's visibility among the rural population as a whole. In this respect, it must be noted that, in parallel with the 1933 administrative reorganization, Godbout changed the workload of agronomists so that they would spend more time in the field, phasing out their traditional activities of giving lectures and demonstrations (see Figure 5.5).[49] Instead, these activities were

FIGURE 5.5 Agronomic Service's activities, annual averages per agronomist, 1920–32

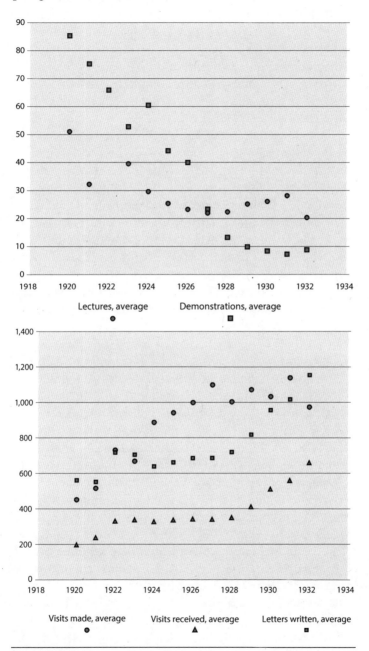

Note: Dissemination activities (top); administrative activities (bottom).
Source: RAMA, respective years.

performed by instructors specializing in specific production sectors.[50] In absolute numbers, the 1933 reorganization allowed agronomists to interact directly with farmers more often, increased the number of lectures, demonstrations, and visits, and imposed a relative ceiling on clerical activities, thereby reversing a trend set in the early 1920s. Since the abrogation of the Agricultural Instruction Act and attainment of the objective of "one county, one agronomist," the agronomists had given, on average, fewer and fewer demonstrations and lectures, and activities related to office work had remained steady (sending newsletters and circulars) or grown slightly (writing letters and receiving visits). A rise in the number of assistant agronomists and secretaries in the county agronomy offices had not checked the trend toward a decline in activities in the field. The presence of this extra personnel, as well as the establishment of offices in every county in the province, was consistent with a growth in the volume of office work. Starting in 1933, the department hoped to reverse this trend, such that its agronomic activity would result in direct contact with farmers, as a new category of employees called "agronomy specialists," with expertise in drainage, poultry production, and horticulture, was introduced.[51] This need for proximity was also integral to the department's reorganization following the retirement of François Narcisse Savoie, director of the Agronomic Service for twenty years, and, especially, once the Union nationale came to power in 1936. Although the new minister of agriculture, Bona Dussault, did not fundamentally reorganize the regional agronomy divisions (a single new district was created), he defined district boundaries so that agronomists could work on a smaller scale and maintain "more frequent and personal contact with farmers."[52]

The formation of agronomy districts independent of the counties – administrative units also used as electoral ridings – was intended to provide a geographic and agricultural basis for the department's local activities. If, in the final analysis, Perron's map had had little influence, it is perhaps because, in comparison to the electoral riding, it limited agronomists' range of action in terms of the department's need for visibility with its clientele. Agronomist Albert Rioux has noted that Caron, minister of agriculture between 1909 and 1929, exerted pressure to make county agronomists act as electoral agents of the Liberal government in rural communities.[53] In this, Rioux is taking up a complaint expressed by political opponents about the Liberal minister in the 1920s.[54] Even after Caron's departure, the chief of the Agronomic Service reiterated the need for agronomists to "refrain from taking their vacation" during the election period in 1931.[55] In fact, by deploying agronomists and forming agronomic districts, the Department of Agriculture wished to highlight its presence and encourage voters to appreciate the action of the government party in power, while at the same time not preventing agronomic activity from improving farmers' conditions, whether they were involved in commercial production or subsistence farming.

Demonstrations and Regional Specialization of Agricultural Production

The deployment of agronomists throughout the province starting in the second decade of the twentieth century enabled the Department of Agriculture to maintain a constant and widespread presence among the rural population. This intervention was only the latest step in the efforts made by the state since the mid-nineteenth century to encourage dissemination of technoscientific innovations and diversification of agricultural production: both the advent of official agronomists and the setting up of specialized services furthered these efforts to reorient agriculture. The dairy industry seemed to be the main beneficiary, but other sectors also profited from the diversification and regional specialization of agriculture orchestrated by the state and its partners.[56] Aside from the annual subsidies paid to county agricultural societies and farmers' clubs, the government provided funding to associations specializing in livestock raising, market gardening, and fruit growing. In the late nineteenth century, the Department of Agriculture became more interventionist by funding the creation of organizations such as the Société d'industrie laitière de la province de Québec.[57] Then, by enlisting technoscientific personnel and adding specialized services to its organization, it internalized the skills needed to structure the agrarian landscape of Quebec, both through dissemination activities and by interventions arising from the experimental enterprise undertaken by the specialized associations.

By increasing its interventions, the department ensured that some kinds of production would be intensified and conferred a sector-based vocation on certain agricultural regions. In fact, this was the purpose of the program, proposed by Perron, undertaken in 1929 by the Department of Agriculture to counter the sparseness and low quality of production. The minister then stated: "Apples should be produced where apples are of an economic and assured value, sheep in sheep districts, milk and dairy cows in regions where the soil and transportation conditions permit. We would like to group the main types of agriculture production in specific locations in the province."[58]

Although the economic crisis of the 1930s put an end to implementation of Perron's reform, which sought explicitly to divide the rural territory into regional agriculture specializations, a *dispositif* had already been used to encourage the adoption of specific techniques and types of production in certain regions. Experimental orchards, poultry stations, and demonstration fields were part of the *dispositif* that the department deployed, through its own agents or through farmers subsidized by the government, to stamp specialized production upon the ecumene as a function of the "vocations" of the soil. With guidance from a few bureaucrats following Édouard-André Barnard, the reorientation of agriculture driven by the Quebec state led to both a rationalization of the rural environment and, ultimately, the imprinting upon the landscape and consequent naturalization of certain kinds of regional production, such as fruit growing and poultry production.

From Arboricultural Stations to Demonstration Orchards: The Regional Concentration of Apple Growing

Fruit growing reached the scale of commercial production in Quebec thanks to expanding urban markets in the second half of the nineteenth century.[59] In the main fruit-growing centres in the province – Abbotsford, Missisquoi, Brome, Châteauguay, Shefford, and L'Islet – fruit growers organized into associations to improve their production techniques and facilitate the marketing of their harvests. At the end of the century, however, they were struggling with competition from early fruits from Ontario and California, which arrived in the Montreal market before their own produce reached maturity. Commercial growers of soft fruits, such as pears, and pitted fruits, such as peaches and plums, were heavily affected.[60] Only apple production seemed to thrive, and the Department of Agriculture supplied financial assistance to associations in this sector for the dissemination of techniques and commercially cultivable varieties. Growers selected apple tree varieties that resisted harsh winters and eliminated those that lacked rusticity. The Pomological and Fruit Growing Society of the Province of Quebec, formed in 1893, offered an annual list of "the best varieties to recommend to apple growers," which, in addition to their climate-related advantages, could be exported and sold profitably in the British market.[61]

With the help of Auguste Dupuis, a nursery owner in Saint-Roch-des-Aulnaies well known to the fruit-growing community, the department funded the creation of six "arboricultural stations" in 1896. Dupuis had expressed his disappointment to the minister that the only arboricultural experiments were taking place at the Central Farm in Ottawa, under conditions that were "of no use to the 'Eastern' region of the province of Quebec, on either shore of the St. Lawrence," and that Nova Scotia had the benefit of a federal experimental farm for the farmers of that province.[62] Created at the instigation of the Pomological Society, the arboricultural stations initially had a five-year mandate, but the department kept them running afterward to use as plant distribution sites.[63] In exchange for an allowance and fruit tree saplings, a superintendent agreed to follow the department's instructions for maintaining an orchard of which he had to make "reasoned exploitation" and that was to be used to supply "information useful to fruit growers of the region."[64] The experiments were aimed at verifying types of crops, sizes of trees, and spraying techniques as a function of the climate and at selecting appropriate varieties. Established in six counties – Beauce, Chicoutimi, Compton, Gaspé, L'Islet, and Maskinongé – chosen with the approval of representatives from the Pomological Society, the stations had a "regional distribution [that] ensured arboricultural experimentation appropriate to the various types of climate in Quebec."[65] However, the department realized that these were not actual horticultural centres, especially in the cases of Chicoutimi, Gaspé, and Maskinongé. In fact, it later had to concede that "the thousands of plantations had not been conclusive in recent decades," especially in sites with difficult climatic conditions, such as Caplan, Chicoutimi, Mistassini, Mont-Louis, and Roberval.[66] Beyond agronomic considerations, the presence of religious communities justified the creation of some stations in zones that were marginal for fruit production, such as the ones in Beauceville, Roberval, and Rimouski.

Another method of development and dissemination of arboricultural techniques emerged in 1911, when the department, again endorsing an idea of the Pomological Society, established demonstration orchards "as a complement to arboricultural stations": two in Rouville County and one each in Huntingdon and Deux-Montagnes counties.[67] Demonstration orchards enabled arboriculture instructors to conduct experiments and pass methods along. They were under the administrative control of a seven-member committee – three members appointed by the Pomological Society, three by the cooperatives, and one by the department. The orchard was supervised by a cooperative of orchard owners who took part in "practical and reasonable" experiments that had to be executed uniformly in all demonstration orchards.[68] The experiments were conducted under the auspices of the department, which established control plots in the orchard and other plots in which particular cultivation and spraying methods or varieties were tested.

These two instruments, arboricultural stations and demonstration orchards, formed the basis of the Horticultural Service that the department created the year after the Agronomic Service was established. The service was involved with the management and inspection of demonstration orchards and fields, which it supplied with plants grown at the Department of Lands and Forests' Berthierville nursery and then, starting in 1918, at the Deschambault nursery, where the Department of Agriculture was now conducting its acclimation tests and raising saplings.[69] Led by Solyme Roy, the service was composed of the manager of arboricultural stations, Auguste Dupuis; the superintendent of demonstration orchards, Peter Reed; and the provincial entomologist hired with an Agricultural Instruction Act grant.[70]

Firmin Létourneau observed that the arboricultural stations and demonstration orchards had different objectives: "In the first case, we are establishing a new industry; in the second, we are perfecting what exists."[71] According to Paul-Louis Martin, each initiative was addressed to different clienteles and helped to divide fruit growers into two classes: smaller producers, most of which were situated in marginal zones, with particular pedoclimatic production conditions, and far from a large market, and "large professional producers that tended toward concentration and specialization," with orchards situated close to major consumption and marketing centres.[72]

An examination of the spatial distribution of the arboricultural stations and demonstration orchards tends to confirm Martin's hypothesis (compare the two maps in Figure 5.6). In fact, these two instruments did have different missions. The arboricultural stations were used to "determine the districts where the best varieties can succeed,"[73] especially when apple tree grafts of different varieties were sent for the purpose of creating new orchards in areas deemed to be propitious to this type of agriculture. The Horticultural Service created nineteen stations between 1914 and 1927; once it felt that "the sites that offer the most advantages for the creation of major fruit-production centres" had been determined, it closed them.[74] In 1925, the year before the decision was made to close the stations, the service stopped operating the demonstration orchards situated in Beauceville, Disraeli, Saint-Léon-le-Grand, Saint-Lin–Laurentides, and Saint-Sylvestre, which, it said, were outside the fruit-growing districts.[75] The following year, the service claimed to have identified the sites that it deemed "conducive to fruit growing" in the province and retained thirty-four orchards "in places where it is possible to grow fruits on a commercial basis." It also enabled religious communities to maintain a few "poorly situated" orchards belonging to teacher colleges "for the purpose of demonstration for students."[76]

How did the arboricultural stations and demonstration orchards contribute to development of the fruit-growing industry in Quebec, notably with regard to yields and to moving production sites? The maps in Figure 5.6 illustrate some characteristics of fruit-growing production in Quebec. In the 1920s, the position of the counties southeast of Montreal as main centres of fruit production in the province was consolidated, to the detriment of counties in the Eastern Townships and Bois-Francs, as well as the Lower

Regionalization and Specialization of Agricultural Production 145

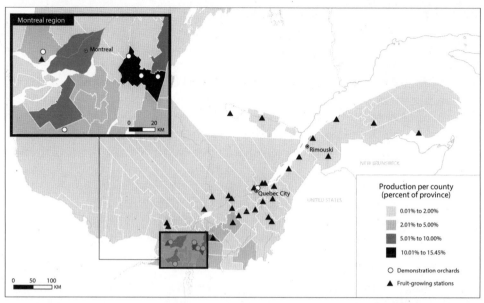

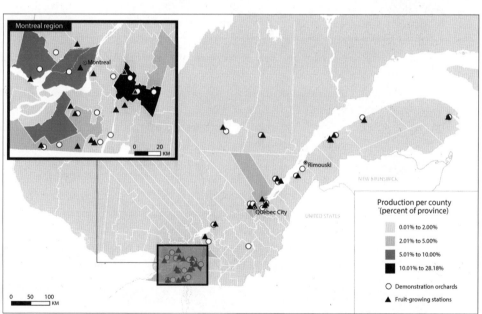

FIGURE 5.6 Apple production per county, 1921, and demonstration sites, 1918 *(top)*; apple production, per county, 1931, and demonstration sites, 1928 *(bottom)*. | Sources: *Recensement du Canada,* respective years; *RAMA,* respective years.

St. Lawrence region, which had been a strong fruit-producing area until the First World War.[77] Only the region around Quebec City seemed to improve its position, even though it could not rival the fruit-growing counties in the Richelieu Valley. This portrait of production is somewhat at odds with the department's efforts. Although several counties remained strongly invested in fruit production (Rimouski, Kamouraska, Bellechasse, and Charlevoix), other regions disappeared from the department's radar, such as the Eastern Townships and the North Shore except for Saint-Maurice County, which retained demonstration orchards. Finally, the counties in which the department had established its first arboricultural stations in 1898 (Beauce, Compton, L'Islet, Gaspé, and Maskinongé) were now stripped of them or simply became marginalized in the state fruit-growing apparatus, with the exception of Chicoutimi, where the religious community initially responsible for planting orchards in the region was still active, unlike that at the Collège de Saint-Joseph-de-Beauce.

In 1927, the Horticultural Service stopped using demonstration orchards as a main dissemination method, preferring "supervised orchards" – commercial orchards whose "owners agreed to leave their management ... to an instructor from the Horticultural Service."[78] This way of working testified to the relative autonomy of the horticultural technicians employed by the department, who no longer had to rely on nursery operators and fruit growers for selection of the sites for their interventions. It is interesting to note the reasons invoked by the department for abandoning the demonstration orchards, which had been in operation since 1909 and had had a decisive effect on adoption of the growing methods promoted by the department's instructors. The department saw this move as a way to get rid of opportunists who "profited more from immediate material benefits than long-term improvements" and whose orchards, "once the contract ended ... fell back into the negligent state that they had previously been in."[79] Nevertheless, rather than "take technical direction of their operation and monitor on site the application of our growing methods that each agreed to follow to the letter," the department opted to intervene in management, "in order to be able to assume greater control of the largest possible number of commercial orchards and ensure their success."[80]

When the department announced the discontinuation of the arboricultural stations and demonstration orchards, it stated that it had delimited the "fruit district," where it planned to install 125 "supervised orchards" in order to "develop our fruit-production centres much more quickly."[81] In fact, this was an action that Perron emphasized when he presented his reform:

FIGURE 5.7 As it organized cooperative services for spraying orchards in the apple-growing regions of the province, the Plant Protection Service conducted tests to establish spraying schedules. The tests also provided an opportunity for demonstrations for the region's producers. | Protection of orchards in Rougemont: spraying with sulfur fungicide, by Paul Carpentier, 1941. *Source:* BANQ-Québec, Fonds ministère de la Culture et des Communications, Office du film du Québec, Documents iconographiques, E6, S7, SS1, P3644.

Thanks to an inventory that we have had prepared over the last six weeks, we know that fruits may be commercially cultivated advantageously and particularly profitably in certain favoured regions of our province ... It is desirable to increase the plantation of commercial orchards in all appropriate regions where well-drained and gravelly soils are the most propitious to fruit cultivation. For this progress to be made, the sites appropriate for fruit cultivation must immediately be defined, farmers must be encouraged to purchase good trees, and they must be monitored constantly so that they do not make mistakes either in their planting or in how they protect their harvests against the ravages of diseases.[82]

Although the department promoted the Montreal and Quebec City regions, it recognized in the latter case that "the cultivation of orchards doesn't have the significance that it has reached around Montreal."[83] The existence of developing centres on Île d'Orléans and in Saint-Nicolas, notably, encouraged Maurice Talbot, orchard inspector at the Horticultural Service, to organize some fifty producers in these two locations. He also began a spraying service in the Montreal and Quebec City districts in 1928 (see Figure 5.7), confirming the regional specialization of apple production in very limited areas of the province.[84] While organization of the Quebec City district remained in abeyance, the department defined three zones near Montreal, including, roughly, the orchards in the counties of Rouville, Huntingdon, and Châteauguay, as well as those in Jacques-Cartier and Deux-Montagnes.[85] In the view of the department, these three zones formed "the perfect fruit-growing district," where apple producers agreed to offer the "varieties in demand" – the Fameuse and the McIntosh – rather than opting for "the detestable habit of trying all apple varieties extolled by skilful advertisements without further thought."[86] Here again, the issue was to encourage the plantation of commercial orchards and regional apple-growing specialization, as the department continued to deplore the constant imports of freight-car loads of apples into the province in great numbers to serve the local market.[87]

From Henhouses to Cooperative Hatcheries: Industrialization of a Domestic Activity

The rise of horticultural production such as apple growing was due to a combination of pedoclimatic conditions, the proximity of urban markets, and the desire to reclaim the local market or address the export market. Less dependent on environmental and geographic factors, poultry production was industrialized on the basis of a series of scientific and technical advances.[88] The upheavals that this change caused on farms, notably with regard to the gendered division of labour and women's position in the domestic economy, were both unexpected and significant. The emergence of this industry must be seen as the result of concerted efforts to have farmers gradually abandon artisanal poultry raising through the introduction of pure breeds and artificial incubation, construction of poultry buildings, distribution of balanced feeds, and organization of egg and poultry marketing.

If the Department of Agriculture played an essential role in this transformation, it was because it was supported by the initiatives of producers and their leaders. Somewhat comparable to the case of fruit production, an association was at the core of development of this agricultural specialty: the Union expérimentale des agriculteurs de Québec. A cooperative for agricultural propaganda and education, launched in 1909 on the model of a similar organization in Ontario and definitively organized in 1911, the Union had a mandate that was general rather than sector-based or geographic, unlike most other agricultural associations at the time.[89] This mandate probably reflected the priorities of its parent institution, the Institut agricole d'Oka. Nevertheless, its aviculture interventions made it a prominent actor in this sector, especially because its founder and secretary, Brother Liguori, was an ardent promoter of poultry production. A teacher at and then director of the Institut agricole d'Oka, Brother Liguori made the Union the main lever for agricultural renewal.[90] He also directed the aviculture bureau, whose headquarters he established at the École normale de Québec, "in the immediate vicinity of the Department of Agriculture and [where] he now became virtually the chief of the provincial Poultry Service."[91]

It must be said that aviculture was of little interest before the late nineteenth century. Chickens were kept in the stable or piggery in the winter and left free during the summer. Poultry production became a high priority for the Department of Agriculture starting in 1910, however, when it conducted a study on the importing of eggs to supply urban markets.[92] For more than a decade, stakeholders in the sector joined forces to intensify commercial poultry production. Among them was the Union expérimentale des agriculteurs, whose aviculture bureau took action to improve the marketing of eggs.[93]

In addition to organizing a regular poultry market in the cities, the Union saw to the erection of henhouses in different counties. These farm buildings, independent of the stables or piggeries where poultry were usually kept, were intended to help farmers control the production of eggs in winter and fatten the animals through the use of meal.[94] To its "experimenter members," numbering 325 in 1912, the Union gave a bonus and donated labour for construction of henhouses that would serve as demonstration sites for the farmers in their county: "And it is the intention of the Union to grant bonuses only in the regions where these types of husbandry and methods are still unknown or almost unknown."[95] When it was founded, upon the suggestion of the Department of Agriculture – from which it received generous funding – the Union established three poultry-fattening stations, in Foster, L'Anse-à-Gilles, and Saint-Pascal; there were twenty-two stations by 1912.[96]

In 1914, the Department of Agriculture established a poultry service that gradually took over the Union's infrastructure, initially by providing the fattening stations with instructors and demonstrators.[97] The ties with the Union's activities were all the stronger because the first director of the Poultry Service at the department was Brother Liguori. A division of labour was set up between the Union and the service, with the former being more focused on experimentation even though it had to answer to its members, who saw it as being too focused on science.[98] The Union used money previously allocated to managers of fattening stations to concentrate on breeding stations and encourage the adoption of purebred flocks and artificial incubation. It continued its experimental activities in the farm building of the agriculture school at the École normale de Québec, where, in 1913, it installed an aviculture station to "test cooperative poultry-production methods: cooperative incubation, cooperative breeding and fattening, etc."[99] With the help of a grant from the federal government, the Department of Agriculture turned this station into an aviculture school to offer practical instruction "from Trois-Rivières to Gaspé." On the instructions of Brother Liguori, the department provided the farm building with a cooperative incubator and a brooder with a capacity of a thousand chickens. The first chickens produced were distributed throughout the province to "propagate good breeds and good lines"[100] and thus to replace mixed-breed poultry. The department had the same equipment installed in its breeding stations; its policy was to let the Union conduct the first tests and then distribute the equipment into ordinary stations if they proved successful.[101]

In addition to educational activities, the Poultry Service ran stations for fattening and incubation. In 1927, the service estimated that it had established 168 aviculture stations since its foundation, although only 54 were still in operation. The fattening stations, of which there were 35 in 1930, were intended to spread rational feeding methods. They were set up "in centres where modern aviculture was not yet known and were located with farmers whose farms were identical to the average."[102] Like the demonstration orchards, they were managed by farmers, who had to prove that they met certain criteria with regard to flock size and equipment in order to receive funding. During the two years that the fattening stations were in operation, they were monitored closely by the aviculture instructors.

The incubation stations were central to the poultry industry development strategy during the interwar period. Following the Union's lead, the department's agents constantly condemned the farmers' propensity to keep poultry of degenerate or worthless breeds, such as those entered into competitions but with no productive value: "It is important to prevent the intensive multiplication of the breeds that are promoted at fairs and that lead farmers to easily procure breeds with no guarantee of constant success."[103] As well as organizing research on egg fertility and feeding of chickens around two poultry breeds – the Barred Plymouth Rock and the Rhode Island Red, the most popular in American hatcheries and poultry houses – the Poultry Service sought to introduce purebred hens and roosters by creating incubation stations throughout the province.[104] Situated in Princeville, Shawville, Saint-Prime, Saint-Joseph d'Alma, Saint-François-du-Lac, and Rimouski, these stations were "intended to expand [the use of] better breeds," for the Poultry Service deemed that "the problem of scarcity of good subjects for breeding remains an obstacle to the development of profitable poultry production."[105]

Although production seems to have been solidly established after several decades of effort by the Poultry Service and the Union, marketing of eggs and poultry still seemed insufficient in the view of stakeholders. To overcome this problem, Perron made cooperation core to his program of agricultural renewal.[106] In June 1929, he explained the objectives that he was assigning to the sector as follows:

> In aviculture ... we have production that should be better developed, especially when we know that our consumption centres import enormous quantities of eggs and poultry. We intend to develop this production in the province. My department will offer all information on the most profitable recognized methods free of charge. We will develop production centres that could interest major purchasers in large cities twelve months of the year. We will organize distribution centres for chicks and [select] good subjects for reproduction and raising.[107]

In the wake of this intervention, the Poultry Service conducted a study on egg and poultry production in Quebec – the second in less than twenty years – and gathered "the provincial aviculture instructors, the federal aviculture promoters ... the professors from agriculture schools, and some agronomists" at a conference to "formulate an aviculture extension and improvement program."[108] The conference led to a five-year program aimed at creating centres for producing selected day-old chicks. The program provided the orientation for the Poultry Service for the following decade as it ceased to fund aviculture stations and turned to organizing poultry production associations through governmental subsidies for cooperative purchasing of incubators. The first association to be created was the Coopérative avicole de la Vallée de la Chaudière by producers in Beauce and Dorchester counties in 1930. After this first experiment failed, the Poultry Service decided to concentrate its activities on cooperative hatcheries, organization of which had begun even before the launch of Perron's plan. There were eight in 1931, thirty-seven in 1935, and forty-five in 1941.[109]

How did the interventions of the Poultry Service impact poultry production and its growth and regional distribution? During the period under study, poultry production rose markedly, first between 1900 and 1910, then in the 1920s; it fell in the 1930s (see Figure 5.8). Between 1901 and 1941, province-wide, certain districts (Lower St. Lawrence, Montreal-Ottawa Valley,

FIGURE 5.8 Contribution by district to provincial poultry production, 1901–41

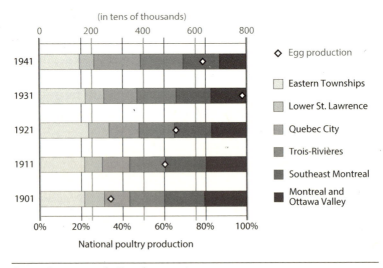

Source: Recensement du Canada, respective years.

southeastern Montreal, and the Eastern Townships, with the exception of Beauce County) fell in importance in provincial production, while others (the Trois-Rivières and Quebec City districts, which also included counties in the northern part of the province) made a greater contribution. The Department of Agriculture supplied the poultry stations in the Quebec City district (particularly those in the counties on the south shore of the river) generously, leading to remarkable rises in production (compare the maps in Figure 5.9). It also favoured the Lower St. Lawrence district, where relative production had declined, unlike the Trois-Rivières district, where the number of stations remained relatively stable (except for a few years), despite a greater contribution to provincial production. Finally, there were a small number of stations in the Eastern Townships and the counties north of Montreal, where production fell relative to that in the province overall, and the number of stations rose slightly. All districts, with the exception of the Eastern Townships, had incubation stations, although the districts around Quebec City (where the first cooperative incubators were formed), Trois-Rivières, and the Montreal-Ottawa Valley had them for a longer time. For the Quebec City and Montreal-Ottawa Valley districts, the stations were situated in educational institutions: the Institut agricole d'Oka, the École de l'industrie laitière in Saint-Hyacinthe, and the École normale de Québec. It was also in these

Regionalization and Specialization of Agricultural Production 153

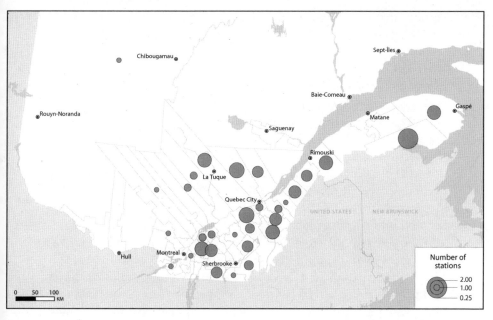

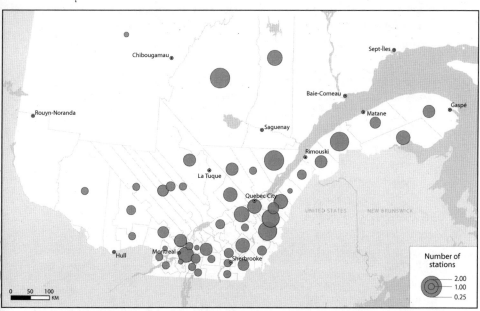

FIGURE 5.9 Poultry stations in operation, by county, 1921 *(top)* and 1931 *(bottom)*. | Sources: *Recensement du Canada,* 1921, 1931; *RAMA,* 1921, 1931.

locations that the Department of Agriculture funded the research effort to consolidate its own resources in the dissemination of knowledge.

FROM THE TACHÉ COMMISSION'S REPORT in the mid-nineteenth century to Minister of Agriculture Joseph-Léonide Perron's reform in 1929, the province's agriculture was profoundly transformed. Starting in 1875, an agricultural renewal followed the rise of dairy production, the generalization of mixed polyculture, and the adoption of specialized production. Although crises persisted from one era to another, it must be emphasized that the solution advocated in each period to solve the problems of the rural world was based on creating closer links between the agro-ecological milieu and local markets. The Department of Agriculture constantly sought to reorient farmers in this direction through education.

Compared with the other three sectors studied, the interventions of the Department of Agriculture involved less the development of knowledge than its dissemination and application as a function of pedoclimatic and commercial conditions. Nevertheless, the inception of university-level training in agricultural science changed the department's potential for action. During the period under study, specialized services set up within the department hired skilled personnel to offer technical support to producers. Such training had previously been the prerogative of independent associations that had developed their own expertise.

In general, from the early agricultural societies to the districts of the Agronomic Service, the territories of agronomic action were defined according to the administrative borders of counties, independent of the agricultural possibilities of the respective area. Despite the vain attempts by some of the department's agents to "depoliticize" the territories of agronomic action, the reorientation of Quebec agriculture in the early decades of the twentieth century led to production of a new agricultural map of the province, characterized by specialization and regionalization of crop and livestock production. Conversely, over time the territorial organization of agronomic action became dependent on transformations in agriculture initiated by the department's activities, reflecting both progress in and specialization of regional production.

It remains difficult to provide a precise assessment of how greatly the Quebec state's interventions impacted the agricultural reorientation of the province, and the role played by the agronomists seems even less easy to determine. The associations in the fruit-growing and poultry sectors were certainly a crucial part of the formation of the specialized agricultural producer. In addition, the efforts by official agronomists channelled the work of these associative forces at the same time as the associations had infiltrated, for a certain period, departmental services. Finally, the Experimental Farms Branch of the federal Department of Agriculture contributed to the province's agricultural renewal.[110] The Quebec state's agronomic activity nevertheless gained autonomy in relation to the associative forces and political constraints that had always defined its territory. Powerful evidence of this was the maintenance of the Agronomic Service by the Quebec Department of Agriculture once the federal financing that had helped to establish it ended. Rather than confine itself to inspection activities, the department expanded its own improvement and experimentation activities while maintaining the educational enterprise at the core of its technoscientific interventions.

CONCLUSION

Knowledge, Power, and Territory

AS A PRODUCER OF KNOWLEDGE, the Quebec state formulated and implemented programs aimed at occupation and transformation of its territory as the basis for its government of natural resources. Scientific and political reorganization of the territory accompanied an expansion and consolidation of state power.

The themes examined here give a good assessment not only of the extent and diversity of the Quebec state's technocscientific interventions in the early twentieth century but also of the depth of its capacity for action. Long considered ineffective and absent from social and economic life in Quebec, the state proved, rather, to be endowed with scientific and technical resources that it sought to consolidate and did not hesitate to use to intervene in several sectors. Repeatedly, it recovered practices and infrastructure from federal institutions and civil society actors to appropriate territories that it wanted to know more about, to shape, or to regulate. It also set up its own institutions to develop and apply technoscientific knowledge in order to frame the exploitation of natural resources. Thus, over the period studied, a fundamental change was the internalization of a technoscientific capacity within the provincial public administration in order to intervene in Quebec economic and social life.

The presence of the state was manifested not only by an increase in the number of civil servants and by more frequent and diversified interventions but also by the types of interventions, which became more and more targeted. Whereas the surveyors employed by the commissioner of Crown lands served timber limit holders, settlers, mining prospectors, and holders of fishing and hunting licences, the subsequent creation of separate departments and specific

Conclusion

scientific services for these sectors led to the hiring of qualified personnel, illustrating the specialization of the public administration and of knowledge related to the government of natural resources. Similarly, although all departments used a single map when they defined timber limits, hunting domains, falls and rapids for water powers, or a clay belt for colonization, the provincial public administration drew multiple administrative divisions on this map and reorganized the space for exploitation of natural resources. The state was involved in distinct territories whose borders were drawn in some cases because of the environment and the resources to be exploited, in other cases because of the institutions (legal, but also technoscientific) and actors engaged in the extraction of resources.

Specialization of the territory and the public administration was paralleled by the growth of institutions of higher learning. The Quebec state drew qualified personnel from these institutions, and even contributed to the training of its civil servants by placing them in schools that it funded or offering them the opportunity to participate in missions and expeditions led by professors from these institutions and members of its scientific services. Therefore, state sciences were doubly autonomous: first, the state became a partner entirely committed to the national technoscientific enterprise; second, as fundamental partners in this enterprise, the scientific services evolved by adopting characteristics similar to those of scholarly institutions, including specialization of knowledge and reproduction of knowledge producers.

The internal development of scientific expertise was one engine of growth for the Quebec state's administrative capacities. Another was the implementation of this expertise to build territories whose regulation fed the state's power. Through the administrative divisions that they created for each resource, and for the regional specializations that they impressed upon the territory for extraction and processing of resources, the scientists imagined and shaped a national space. Far from grappling with a vast, undifferentiated territory, the public administration defined a series of spaces according to the resources made visible by scientific knowledge and political intervention. The territory was organized in distinct ways in response to the efforts of scientists, the requirements of industry, and the interventionism of a state that, albeit neither omnipresent nor omnipotent, infiltrated the interstices of spaces that its technoscientific personnel were constantly reconstructing.

In seeking to understand the role of technosciences in the formation of the state and the deployment of state power, my goal has been to highlight the means by which representations resulting from scientific activity marked the fields of knowledge and power. One path led me to uncover the processes of making visible different objects of knowledge and government through various spatial practices. Cartography and soil classification, explorations, inventories and surveys, inspection and surveillance, and dissemination of techniques shaped territories for governing populations and resources, for recording and surveying them, and for ordering and exploiting them. Through abstraction and normalization, state agents revealed the riches and properties of territories, as materialized by maps, inventories, surveys, and reports. The information generated by the tools of knowledge made visible and legible resources that were otherwise buried in the soil and subsoil, fleeting or hidden by forests and watercourses, or abandoned on the land or around farm buildings, and that might be otherwise be depleted or never commodified. It also made it possible to assess the exhaustion, potential or real, of resources so that limit holders and property owners could be held liable for abuses perpetrated against the land and its resources, or for their negligence in improving them. In this way, it outlined the obligations that limit holders and property owners had toward the natural resources for which they were responsible and that they had to agree to husband, at the risk of having the state penalize them or force them to give them up. Knowledge delimited the relationship that users had to maintain with the resources in a given territory as a function of norms defined by the state's scientists with regard to how the national riches were to be exploited and the territory occupied. The work of state technoscientific agents made it possible to improve territories and make them productive through the formation of subjects around natural resources and environments that science problematized and politics mobilized.

Through its scientific production, the Quebec state thus was able to create territorialities within and through which it exercised the government of natural resources. In this regard, colonization and resettlement of the hinterland was a recurrent concern, and in this book I have addressed this issue only transversally and superficially, as it was shaped in part by state scientific activity, but also by various institutional and ideological considerations. Similarly, the specialization of scientific services and the growth of the bureaucracy led to further spatial divisions of the province that put not only the borders but also functionalities of certain territories in opposition to the policies of other departments. Although they were a driving force in modernization of the state's intervention mechanisms, scientists had to shape and negotiate a series of spaces within the public administration, including their own space, within which science and politics had to coexist.

APPENDIX

Identification of Technoscientific Activities in the Public Accounts (1896–1940)

MY SYSTEMATIC ANALYSIS of the public accounts between 1896 and 1940 was aimed at identifying all technoscientific activities for which the government made an expenditure. My compilation of expenditures was based on the categorization of technoscientific activities into three enterprises. The *educational enterprise* includes training and dissemination activities. Training was based on direct funding of educational institutions and the granting of scholarships and teaching chairs. Dissemination refers to popularization of technoscientific knowledge through publications, lectures, and extension activities. Intended mainly for operators, dissemination differed from training, which was dispensed in the formal framework of educational institutions. The *descriptive enterprise* involved measurement (laboratory analyses and land surveys) and inventorying (the collection and identification of botanical, zoological, and mineralogical specimens). These activities were based on exploration and the recording of information needed for cartography. Other activities – studies and applications – were aimed at improving the environment and knowledge of the conditions for its exploitation. These comprised the *experimental enterprise*, designated as such to express their interventionist nature, as they were aimed at modifying the environment in order to learn more about it or increase its productivity. Studies involved investigations and research, whereas applications involved regulation of natural resource exploitation practices. Sometimes, applications meant inspections in which duly trained agents ensured that knowledge was properly applied by operators, notably by ensuring that the guidelines that they helped to formulate were respected. Sometimes, they referred to the work underpinning the availability of materials produced by the state in its facilities, such as tree nurseries and fish hatcheries, to shape the landscape and encourage a specific way of exploiting natural resources. Such material production was based on techniques and knowledge regarding the cultivation of plants or the breeding of animals and the maintenance of living material, as well as knowledge of the ecological and economic environments where they were to be deployed.

Although the public accounts gave a preliminary overview of the work executed by the technoscientific personnel within departments, the use of this source is not foolproof. First, data from the public accounts represent only a fraction of the technoscientific expenditures undertaken in Quebec over this period by all actors: I could add, for example, federal agencies, educational institutions, civil associations, and some companies, particularly those associated with the new industrialism. Second, these data are often incomplete. Ruth Dupré has given an exhaustive critique of them in which she notes the rhetorical aspect of the budget statement, although she still uses these data – once they have been standardized – to demonstrate Quebec state interventionism in the first half of the twentieth century. In addition, many expenditures were erased from the public accounts for various reasons, ranging from the position of a good government's responsible budget to creative accounting practices of the time. Dupré speaks, for instance, of the propensity to present the public accounts on a net basis and to manipulate data "to avoid politically embarrassing current deficits."[1] She therefore points to the need to undertake a detailed breakdown of expenditures in the public accounts and to trace expenditures in relation to the function of the intervention funded – which, in my analysis, corresponds to a technoscientific activity. It should also be noted that the indexations intended to present expenditures in constant dollars over the long term rarely offer an entirely satisfactory solution. For my study, I therefore chose to present the relative size of these expenditures and to emphasize orders of magnitude, rather than to give the dollar amounts spent.

Notes

Foreword: Science in Action

I thank Edward Jones-Imhotep for casting an eye over a draft of this chapter, Stéphane Castonguay for his warm endorsement of this attempt to "situate" his groundbreaking work (as well as some small but important corrections to the earlier version), and Suzanne Zeller (whose grasp of the literatures discussed here far exceeds my own) for her careful reading and helpful suggestions for the improvement of my initial efforts.

1. Edward Jones-Imhotep and Tina Adcock, eds., *Made Modern: Science and Technology in Canadian History* (Vancouver: UBC Press, 2018), especially "Introduction," 3–37.
2. For introductions to the extensive literature here, see Anthony Giddens and Christopher Pierson, *Conversations with Anthony Giddens: Making Sense of Modernity* (Cambridge: Polity Press, 1998); Marshall Berman, *All That Is Solid Melts into Air: The Experience of Modernity* (New York: Viking Penguin, 1988); P. Osborne, Michael Payne, and Jessica Rae Barbera, "Modernity," in *A Dictionary of Cultural and Critical Theory*, ed. Michael Payne (West Sussex, UK: Wiley-Blackwell, 2010), 456–59; and "AHR Roundtable: Historians and the Question of 'Modernity,'" *American Historical Review* 116, 3 (2011): 631–751.
3. So the first substantive chapter of Jones-Imhotep and Adcock, *Made Modern*, instantiates the argument of this paragraph: Ephram Sera-Shriar, "Civilizing the Natives: Richard King and His Ethnographic Writings on Indigenous Northerners," 39–59; see also chapter 10 of the same volume, Stephen Bocking, "Landscapes of Science in Canada: Modernity and Disruption," 251–78, for a more general reflection on the theme.
4. For a recent, innovative reflection on the latter point, see Kevin Hutchings, *Transatlantic Upper Canada Portraits in Literature, Land, and British-Indigenous Relations* (Montreal and Kingston: McGill-Queen's University Press, 2020).
5. Jones-Imhotep and Adcock, *Made Modern*, 8.
6. For a useful, broad-ranging examination (and impressive bibliography) reviewing the connections and synergies between environmental history and the history of technology (and science), see Sara Pritchard, "Toward an Environmental History of Technology," in *The Oxford Handbook of Environmental History*, ed. Andrew C. Isenberg (Oxford and New York: Oxford University Press, 2014), 227–58.

7 This phrasing closely follows John V. Pickstone, *Ways of Knowing: A New History of Science, Technology and Medicine* (Chicago: University of Chicago Press, 2001), 8. These ways of knowing are ideal types in the sense developed by Max Weber, which is to say that they emphasize characteristics common to most cases rather than define all facets of any particular instance.
8 See Crosbie Smith, "The Science of Energy: A Cultural History of Energy Physics," in *Victorian Britain* (Chicago: University of Chicago Press, 1999). I am indebted to Suzanne Zeller for this reference.
9 Pickstone, *Ways of Knowing*, 14.
10 For cogent earlier summaries of the Canadian story here, see Zeller, *Land of Promise, Promised Land*; C.C. Berger, *Science, God and Nature in Victorian Canada* (Toronto: University of Toronto Press, 1983). Suzanne Zeller, "Canada," in *The Cambridge History of Science*, vol. 8, *Modern Science in National, Transnational and Global Context*, ed. Hugh R. Slotten, Ronald L. Numbers, and David N. Livingston (Cambridge and New York: Cambridge University Press, 2020), 736–51, is a thoughtful, wide-ranging, and indispensable treatment of these matters that also draws upon Pickstone, *Ways of Knowing*.
11 P. Henry Gosse, *The Canadian Naturalist: A Series of Conversations on the Natural History of Lower Canada* (London: John van Voorst, 1840); Ann Thwaite, *Glimpses of the Wonderful: The Life of Philip Henry Gosse, 1810–1888* (London: Faber and Faber, 2002), 67.
12 Winifred (Cairns) Wake, "Prince Edward Island's Early Natural History Society," *Island Magazine* 37 (Spring/Summer 1995): 27–33; see also for earlier, individual interest, Elinor Vass, "Mrs. Haviland's Plants," *Island Magazine* 36 (Fall-Winter, 1994): 23–25. For an overview, see Berger, *Science, God and Nature*.
13 C. Gordon Winder, "Logan, Sir William Edmond," in *Dictionary of Canadian Biography*, vol. 10 (Quebec City/Toronto: Université Laval/University of Toronto, 2003), http://www.biographi.ca/en/bio/logan_william_edmond_10E.html; Zeller, *Inventing Canada*.
14 Zaslow, *Reading the Rocks*.
15 For more on Logan, see Suzanne Zeller, "Geology's Extractive Impulse," in *Extraction Empire: Undermining the Systems, States, and Scales of Canada's Global Resource Empire*, ed. Pierre Belanger (Cambridge, MA: MIT Press, 2018), 234–49.
16 Ingram, *Wildlife, Conservation, and Conflict in Quebec*, 66–67.
17 A.B. McCullough, "Wilmot, Samuel," in *Dictionary of Canadian Biography*, vol. 12 (Quebec City/Toronto: Université Laval/University of Toronto, 2003), http://www.biographi.ca/en/bio/wilmot_samuel_12E.html; Knight, "Samuel Wilmot, Fish Culture, and Recreational Fisheries in Late 19th Century Ontario."
18 Canada, "Report of Fish-Breeding in the Dominion of Canada, 1877," *Sessional Papers*, No. 1 (Ottawa: 1878), 24, cited in Knight, "Samuel Wilmot, Fish Culture, and Recreational Fisheries in Late 19th Century Ontario," 80, which reproduces a panoramic view of the Newcastle hatchery.
19 Pickstone, *Ways of Knowing*, 15, 14; the original uses the plural "intertwinings."
20 Stéphane Castonguay, "Naturalizing Federalism: Insect Outbreaks and the Centralization of Entomological Research in Canada 1884–1914," *Canadian Historical Review* 85, 1 (2004): 1–34.
21 Suzanne Zeller, "Roads Not Taken: Victorian Science, Technical Education, and Canadian Schools, 1844–1913," *Historical Studies in Education/Revue d'histoire de l'education* 12, 1 (2000): 1–28. See also Suzanne Zeller, "'Merchants of Light': The Culture of Science in Daniel Wilson's Ontario, 1853–1892," in *Thinking Hands*, ed. Elizabeth Hulse (Toronto: University of Toronto Press, 1999), 117–23.
22 Jacques Saint-Pierre, "L'École d'agriculture de Sainte-Anne-de-la-Pocatière," http://encyclobec.ca/communaute_religieuse.php?idcommu_religieuse=36&theme=eduquer; M.-J. Lettre, "L'enseignement agricole à Sainte-Anne-de-la-Pocatière: une histoire à cultiver," *Histoire Québec* 21, 2 (2015): 5–9.

23 Canada, *The Dominion Experimental Farms* (Ottawa: Ministry of Agriculture, 1925). By 1911 there were nine such establishments, and in the mid-1920s the Experimental Farms System comprised "the Central Experimental Farm at Ottawa, twenty-two Branch Farms and Stations, one Tobacco Experimental Station, one Horse Breeding Station, and eight Experimental Sub-stations."

24 G.M. Dawson, "The Progress and Trend of Scientific Investigation in Canada," *Royal Society of Canada, Proceedings and Transactions* 12 (1894): lxvi, quoted in Suzanne Zeller, "Darwin Meets the Engineers: Scientizing the Forest at McGill University," *Environmental History* 6, 3 (2001): 429.

25 Zeller, "Roads Not Taken," 14.

26 "The University and the Manufacturer," *Industrial Canada* 6, 3 (March 1905): 478–79, cited in Zeller, "Roads Not Taken," 15.

27 Zeller, "Roads Not Taken," 14.

28 C.J.S Warrington and R.V.V Nicholls, eds., *A History of Chemistry in Canada* (Toronto: Isaac Pitman and Sons, 1949), quote in R.P. Graham, "Review of *A History of Chemistry in Canada*," *Journal of Chemical Education* 27, 2 (1950): 110. For a more encompassing perspective on these developments, see Zeller, "Canada."

29 Allan Greer and Ian Radforth, eds., *Colonial Leviathan: State Formation in Mid-Nineteenth Century Canada* (Toronto: University of Toronto Press, 1992), 10.

30 Alexander B. Murphy, "Entente Territorial: Sack and Raffestin on Territoriality," *Environment and Planning D: Society and Space* 30, 1 (2012): 161–62.

31 Juliet J. Fall, "Lost Geographers: Power Games and the Circulation of Ideas within Francophone Political Geographies," *Progress in Human Geography* 31 (2007): 195–216.

32 Murphy, "Entente Territorial," 160, 161.

33 See Juliet J. Fall, "Reading Claude Raffestin: Pathways for a Critical Biography," *Environment and Planning D: Society and Space*, 30, 1 (2012): 173–89.

34 Murphy, "Entente Territorial," 163–65, drawing comparison between Raffenstin's formulation and that by Jennifer Wolch and Michael Dear, *The Power of Geography: How Territory Shapes Social Life* (Boston: Unwin Hyman, 1989), who use the "geographically organized human activity" phrase on p. vi, and Francisco R. Klauser, "Thinking through Territoriality: Introducing Claude Raffestin to Anglophone Sociospatial Theory," *Environment and Planning D: Society and Space* 30, 1 (2012): 106–20.

35 My discussion of Sack and Raffestin draws heavily from Murphy's excellent discussion in "Entente Territorial," 159–72.

36 Later published as Thongchai Winichakul, *Siam Mapped: A History of the Geo-Body of a Nation* (Honolulu: University of Hawaii Press, 1994). Benedict Anderson subsequently credited this dissertation with changing his understanding of the development of nationalism in a new chapter, "Census, Map, Museum," included in the second edition of his *Imagined Communities: Reflections on the Origin and Spread of Nationalism*. See also Terence Chong, "Nationalism in Southeast Asia: Revisiting Kahin, Roff, and Anderson," Review Essay, *Sojourn: Journal of Social Issues in Southeast Asia* 24, 1 (April 2009): 1–17.

37 Jocelyn Létourneau, *La condition québécoise: une histoire dépaysante* (Quebec: Les éditions du Septentrion, 2020).

38 Jones-Imhotep and Adcock, *Made Modern*, 4.

Introduction

1 On the use of science by states, see Pestre, *Science, argent et politique*; Mukerji, *A Fragile Power*; Gibbons et al., *The New Production of Knowledge*. As far back as the seventeenth century, Colbertian France was surveying its forests for naval construction, and in the eighteenth

century forestry management was integrated into Prussian cameralism. See Lowood, "The Calculating Forester"; Mukerji, "The Great Forestry Survey of 1669–1671."
2 Pickstone, "Les révolutions analytiques et les synthèses du modernisme," 40–41; Stafford, "Geological Surveys, Mineral Discoveries, and British Expansion, 1835–1871"; Dupree, *Science in the Federal Government*, 46.
3 Zaslow, *Reading the Rocks*.
4 Jarrell, *The Cold Light of Dawn*; Beaud and Prévost, "Models for Recording Age in Pre-1867 Canada"; Curtis, *The Politics of Population*.
5 For different examples of these scientific enterprises in the Province of Canada, see Zeller, *Land of Promise, Promised Land;* Gaudreau, *Les récoltes des forêts publiques au Québec et en Ontario;* Knight, "Samuel Wilmot, Fish Culture, and Recreational Fisheries in Late 19th Century Ontario"; N. Perron, *L'État et le changement agricole dans Charlevoix*.
6 De Vecchi's doctoral dissertation, "Science and Government in Nineteenth-Century Canada" (University of Toronto, 1978), is still essential reading today. See also his articles published posthumously: De Vecchi, "The Dawning of a National Scientific Community in Canada, 1878–1896," *HSTC Bulletin: Journal of the History of Canadian Science, Technology and Medicine* 26 (1984): 32–58; "Science and Scientists in Government, 1878–1896 – Part I," *HSTC Bulletin: Journal of the History of Canadian Science, Technology and Medicine* 27 (1984): 112–42; "Science and Scientists in Government, 1878–1896 – Part II," *Scientia Canadensis* 29 (1985): 97–113.
7 A.S. Sibbald, "The Civil Service as a Career for a University Graduate," *University of Toronto Monthly* 11 (1910–11): 166–72. Quoted in De Vecchi, "Science and Government in Nineteenth-Century Canada," 335.
8 Hodgetts et al., *The Biography of an Institution*.
9 Popularized by anthropologist of science Bruno Latour, the term "technoscience" (and the adjective "technoscientific") describes the interlacing of science and technique, both in artifacts and in the networking of actors. See Latour, *Science in Action*.
10 Gow, *Histoire de l'administration publique québécoise 1867–1970*.
11 *Dispositif* (sometimes translated as apparatus) refers to a heterogeneous network of discursive, institutional, and material elements that "functioned to define and to regulate targets constituted through a mixed economy of power and knowledge." Rabinow, *Anthropos Today,* 52.
12 For a critique of these approaches, see Driver, "Political Geography and State Formation"; Johnston, "Out of the Moribund Backwater"; Painter and Jeffrey, *Political Geography*. In his book first published in 1952 and reprinted in 2007, *La politique des États et leur géographie*, Jean Gottman took a different approach, although his work was not systematically pursued.
13 Rose, O'Malley, and Valverde, "Governmentality," 87.
14 Hannah, *Governmentality and the Mastery of Territory in Nineteenth-Century America*.
15 Brenner, Jessop, Jones, and MacLeod, "Introduction: State Space in Question," in Brenner et al., *State/Space*, 1–26.
16 Mann, "The Autonomous Power of the State"; Lefebvre, *De l'État*, 259–324; Alliès, *L'invention du territoire*.
17 Cox, *Political Geography*, 6.
18 The complementarity is proposed in Murphy, "Entente Territorial," 157–72.
19 Explained for the first time in Sack, "Human Territoriality," this approach receives a fuller articulation in Sack's *Human Territoriality*.
20 The *locus classicus* is Claude Raffestin's *Pour une géographie du pouvoir*, but see also his "Territorialité" and "Space, Territory, and Territoriality."
21 Häkli, "Territoriality and the Rise of the Modern State"; Paasi, "Bounded Spaces in a 'Borderless World.'"

22 Johnston, "Territoriality and the State"; Cox, "Political Geography and the Territorial."
23 See, among others, Sahlins, *Boundaries: The Making of France and Spain in the Pyrenees*; Leuenberger and Schnell, "The Politics of Maps."
24 Harley, "Maps, Knowledge and Power," and the essays collected in Harley, *The New Nature of Maps*; Wood and Fels, *The Natures of Maps*. For Lower Canada, see Boudreau, *La cartographie au Québec*.
25 Winichakul, *Siam Mapped*, 131.
26 Edney, *Mapping an Empire*.
27 B. Anderson, *Imagined Communities*, 163–85; Craib, *Cartographic Mexico*; Schulten, *Mapping the Nation*.
28 Hannah, *Governmentality and the Mastery of Territory in Nineteenth-Century America*; Carter, *The Road to Botany Bay*.
29 Kirsch, "John Wesley Powell and the Mapping of the Colorado Plateau"; Braun, "Producing Vertical Territory"; Ingold, "Naming and Mapping National Resources in Italy."
30 Vandergeest and Peluso, "Territorialization and State Power in Thailand"; Sivaramakrishnan, "A Limited Forest Conservancy in Southwest Bengal."
31 Escolar, "Exploration, cartographie et modernisation du pouvoir de l'État"; Biggs, "Putting the State on the Map"; Strandsbjerg, "The Cartographic Production of Territorial Space."
32 Foucault, *Sécurité, territoire, population*, 100; Séglard, "Foucault et le problème du gouvernement"; Mukerji, *Territorial Ambitions and the Gardens of Versailles*; Carroll, *Science, Culture, and Modern State Formation*; S. Castonguay and Kinsey, "The Nature of the Liberal Order Framework."
33 James C. Scott, "Nature and Space: State Projects of Legibility and Simplification," in Scott, *Seeing Like a State*, 11–52.
34 Peluso, *Rich Forest, Poor People*; Guha, *The Unquiet Woods*; Jacoby, *Crimes against Nature*; Warren, *The Hunter's Game*; Sandlos, *Hunters at the Margin*.
35 Douglas, *How Institutions Think*; Mukerji, "Intelligent Uses of Engineering and the Legitimacy of Power."
36 Mukerji, "The Political Mobilization of Nature in Seventeenth-Century French Formal Gardens."
37 Jacques Donzelot, "The Poverty of Political Culture," *Ideology and Consciousness* 5 (1979): 73–86, quoted in Rose, O'Malley, and Valverde, "Governmentality," 88.
38 Scott, "Nature and Space," 309–41.
39 Michel Foucault, *Security, Territory, Population: Lectures at the College De France, 1977–78*, ed. Michael Senellart, trans. Graham Burchell (New York: Palgrave Macmillan, 2009), 85.
40 Rose, O'Malley, and Valverde, "Governmentality," 87.
41 Stuart Elden and Jeremy W. Crampton, "Space, Knowledge and Power: Foucault and Geography," in Elden and Crampton, *Space, Knowledge and Power*, 1–16.
42 Elden, "Governmentality, Calculation, Territory," 575; see also Elden, "Rethinking Governmentality."
43 In contrast, in urban history there is a literature attentive to the spatial dimensions of liberal governmentality. See, for example, Otter, *The Victorian Eye*; Joyce, *The Rule of Freedom*.
44 Legg, "Foucault's Population Geographies"; Huxley, "Spatial Rationalities"; Huxley, "Geographies of Governmentality"; Rose-Redwood, "Governmentality, Geography and the Geo-Coded World"; Murdoch and Ward, "Governmentality and Territoriality."
45 Foucault, *Security, Territory, Population*, 21.
46 Ibid.
47 Elden and Crampton, *Space, Knowledge and Power*.
48 Whatmore, *Hybrid Geographies*. On Latourian epistemology, see Latour, *We Have Never Been Modern*.

49 Among jurists, see, in particular, Brun, "Le territoire du Québec"; Immarigeon, "Les frontières du Québec." The geographic literature is certainly richer, and aside from synthetic works – Blanchard, *Le Canada français;* Courville, *Quebec* – see the books in two collections: Atlas historique du Québec (Presses de l'Université Laval) and the regional history syntheses published by the Institut québécois de recherche sur la culture.
50 Courville, *Quebec,* 176–77.
51 Linteau, Durocher, and Robert, *Quebec,* 13–15.
52 Dorion and Lacasse, *Le Québec,* 16.
53 See, for example, Séguin, *La Conquête du sol au XIXe siècle;* Bouchard, *Quelques Arpents d'Amérique;* Saint-Hilaire, *Peuplement et dynamique migratoire au Saguenay;* N. Perron, *L'État et le changement agricole dans Charlevoix;* Little, *Colonialism, Nationalism and Capitalism.*
54 Boudreau, Courville, and Séguin, *Le territoire.*
55 An exhaustive review is given in S. Castonguay, "Faire du Québec un objet de l'histoire environnementale."
56 Charland, *Les pâtes et papiers au Québec;* Hardy and Séguin, *Forêt et société en Mauricie;* Van Horssen, *A Town Called Asbestos;* Gaudreau, *L'histoire des mineurs du Nord ontarien et québécois;* Fortier, *Villes industrielles planifiées.*
57 Martin, *La chasse au Québec.*
58 Massell, *Amassing Power;* Bellavance, *Shawinigan Water and Power;* Vallières, "Les entreprises minières québécoises et la grande dépression des années 1930."
59 For British Columbia, a series of geographic works address the relationships among spatiality, natural resources, and state sciences, but here again they consider mines, forestry, and wildlife separately. Braun, "Producing Vertical Territory"; Rossiter, "Producing Provincial Space"; Peyton, "Imbricated Geographies of Conservation and Consumption in the Stikine Plateau"; Brownstein, "Spasmodic Research as Executive Duties Permit." Two recent doctoral dissertations make use of a similar perspective in studies of the development of geology and agricultural sciences across Canada: Grek-Martin, "Making Settler Space"; P.G. Anderson, "Field Experiments."
60 Linteau, Durocher, and Robert, *Quebec,* 13–15, passim. For the period before 1897, see Hamelin and Roby, *Histoire économique du Québec.*
61 On the notion of "natural riches," see Cooper, "'The Possibilities of the Land.'"
62 Bernard Vigod gives a good overview of the industrial policies of Liberal governments after 1900 in *Quebec before Duplessis.* See also Ryan, *The Clergy and Economic Growth in Quebec;* Pontbriand, *Lomer Gouin, entre libéralisme et nationalisme.*
63 These debates are discussed in Roby, *Les Québécois et les investissements américains.*
64 With regard to political figures in the period covered by this study, on premiers and their political personnel, see Trépanier, *Siméon Le Sage;* M.-A. Perron, *Un grand éducateur agricole;* Vigod, *Quebec before Duplessis;* Gallichan, *Honoré Mercier;* Dutil, *L'avocat du diable;* Genest, *Godbout;* R. Castonguay, *Rodolphe Lemieux et le Parti libéral;* Pontbriand, *Lomer Gouin.* On the economic history of Quebec and comparisons with Ontario, see, among others, Marr and Paterson, *Canada,* 445; R. Armstrong, *Structure and Change;* McRoberts and Postgate, *Développement et modernisation du Québec,* 80. Ruth Dupré performs a systematic and effective critique of these books in "Un siècle de finances publiques québécoises."
65 Bruce Curtis, *Ruling by Schooling Quebec: Conquest to Liberal Governmentality – A Historical Sociology* (Toronto: University of Toronto Press, 2012), 16.
66 On liberalism and its history, see, among others, Rosanvallon, *Le libéralisme économique;* Audard, *Qu'est-ce que le libéralisme?*
67 McDonald, "The Quest for 'Modern Administration'"; Anholt, "An Administrative History of the British Columbia Government Agents"; Hodgetts, *From Arm's Length to Hands On.*

68 Hodgetts et al., *The Biography of an Institution*.
69 Dupré, "The Evolution of Quebec Government Expenditures."
70 On "effects of power" and *dispositif* (apparatus), see Foucault, *Discipline and Punish*, 195–228.
71 C. Armstrong and Nelles, "Contrasting Development of the Hydro-Electric Industry in the Montreal and Toronto Regions."
72 Nelles, *The Politics of Development*, 108–53.
73 Lambert and Pross, *Renewing Nature's Wealth*.
74 S. Castonguay, "Foresterie scientifique et reforestation."
75 Martin, *La chasse au Québec;* Hodgetts, *From Arm's Length to Hands On*, 67–70.
76 Borrowed from Charles Rosenberg, the expression "institutional ecology" denotes the relations among research facilities (universities, industrial laboratories, departments of different governmental levels) instituting the division of labour that underlies the development of scientific disciplines. See Rosenberg, "Toward an Ecology of Knowledge."
77 This approach, based on analysis of the relative autonomy of state institutions, is inspired in part by works on historical sociology of the state by Evans, Rueschemeyer, and Skocpol, *Bringing the State Back In;* Rueschemeyer and Skocpol, *State, Social Knowledge, and the Origins of Modern Social Policies*.
78 S. Castonguay, *Protection des cultures, construction de la nature;* Doern, *The National Research Council in the Innovation Policy Era;* Senate Special Committee on Science Policy, *A Science Policy for Canada*, vol. 1 (Ottawa: Queen's Printer for Canada, 1970).
79 *Notre milieu; L'Agriculture; La Forêt; Pêche et chasse*. The inventory also included work on the mines that were published in volumes 20 and 21 of *L'Actualité économique* (1945–46). On the inventory, see Foisy-Geoffroy, *Esdras Minville*, 106–9.
80 Chartrand, Duchesne, and Gingras, *Histoire des sciences au Québec*, 284–96.
81 Two sectors that were important during the period under study, hydroelectricity and the maritime fishery, are not addressed in this book. Compared with these sectors, the forestry, mining, wildlife, and agricultural sectors concern resources the exploitation of which was regulated by access to land. As a consequence, they constitute a privileged field for comparative and transversal study of the governmental interventions by the Quebec government, and more particularly from the perspective of production of territorialities. In addition, Claude Bellavance and David Massel have largely covered the development of the Commission des eaux courantes du Québec and the Service hydraulique, the governmental institutions responsible for supervising, among other things, the exploitation of hydroelectric resources. See Bellavance, "L'État, la 'Houille blanche' et le grand capital"; Massell, *Amassing Power*, and, more recently, *Quebec Hydropolitics*. Similarly, for ocean fisheries, Jacques Saint-Pierre has studied the scientific work undertaken at the Station biologique de l'Université Laval and the École supérieure des pêcheries de Saint-Anne-de-la-Pocatière in his book *Les chercheurs de la mer*.

CHAPTER 1: THE ADMINISTRATIVE CAPACITIES OF THE QUEBEC STATE

1 Gow, *Histoire de l'administration publique québécoise*, 64–71.
2 Linteau, Durocher, and Robert, *Quebec*, 74–95.
3 Gow, *Histoire de l'administration publique québécoise*, 159, 161.
4 On the scope of these administrative changes, see McGee, *Le département des Terres de la Couronne*.
5 Linteau, Durocher, and Robert, *Quebec*, 91–95; Vallières, "Le gouvernement du Québec et les milieux financiers de 1867 à 1920."

6 *Statuts de la province de Québec*, 2 Geo. V (1912), c. 11, art. 6. Quoted in Gow "L'histoire de l'administration publique québécoise," 399 (our translation).
7 Bachand, *Histoire de l'École de laiterie de Saint-Hyacinthe;* Gélinas, *L'enseignement et la recherche en foresterie.*
8 S. Castonguay, "Foresterie scientifique et reforestation."
9 Among others, there were *Statuts de la province de Québec*, 5 Éd. VII (1905), c. 17; 1 Geo. V (1912), c. 16; 11 Geo. V (1921), c. 37. These statutes covered, respectively, obligatory inspection of factories, certification of factories, and integration of inspectors with the Department of Agriculture. Dupré, "Regulating the Quebec Dairy Industry," 346. On the regulatory framework for the dairy sector, see also Otis, "La différenciation des producteurs laitiers et le marché de Montréal."
10 In the specific case of the Department of Lands and Forests, inspection refers here to inventorying and collecting information on the territory to separate land for forestry operations from land for agricultural uses.
11 *Rapport annuel du ministère des Terres et Forêts* (*RAMTF*), 1909–10, 51.
12 *RAMTF,* 1929–30, 516–20.
13 *Rapport annuel du ministère de la Colonisation et des Pêcheries* (*RAMCP*), 1929–30, 514.
14 This statute replaced one that had been adopted a year earlier, the Agriculture Aid Act of 1912 (2 Geo. V, c. 3), which had allowed grants of $500,000 to the provinces. See Fowke, *Canadian Agricultural Policy,* 246–47.
15 Maxwell, *Federal Subsidies to the Provincial Governments in Canada,* 200.
16 Before starting to give university-level courses, the schools at La Pocatière and Oka – founded, respectively, in 1859 and 1893 – offered short courses to the sons of farmers to teach them the rudiments of the trade. Lévesque et al., *150 ans d'enseignement agricole à La Pocatière;* Père Louis-Marie, *L'Institut d'Oka;* Snell, *Macdonald College of McGill University.*
17 Pepin, *Histoire et petites histoires des vétérinaires du Québec;* Bibliothèque et Archives nationales du Québec (BANQ), Québec (BANQ-Québec), Fonds ministère des Terres et Forêts, correspondance générale E21, S 10 (now BANQ-Québec, E21, S 10), versement 1960/01-038/559, lettre 18884/10, G.C. Piché au ministre des Terres et Forêts, 3 juin 1910.
18 *Rapport annuel du ministère de l'Agriculture* (*RAMA*), 1912–13, v.
19 *Statutes of Canada,* 3–4 Geo. V (1913), c. 5.
20 Gow, *Histoire de l'administration publique québécoise,* 50.
21 Archives du Séminaire de Québec (ASQ), Université/61/106, Gustave Piché to the Minister of Lands and Forests [spring 1909]; S. Castonguay, "Foresterie scientifique et reforestation."
22 On Obalski, see Marc Vallières, "Obalski, Joseph," in *Dictionary of Canadian Biography,* vol. 14 (Quebec City/Toronto: Université Laval/University of Toronto, 2003), http://www.biographi.ca/en/bio/obalski_joseph_14E.html. On the early days of the Geological Survey of Canada, see Zaslow, *Reading the Rocks.*
23 S. Castonguay, "Connaissance et contrôle du milieu," 295–307.
24 Ingram, *Wildlife, Conservation, and Conflict in Quebec.*
25 *Statuts refondus du Québec,* 48 Vict. (1884) c.12, art. 5493.
26 *Rapport annuel du ministère de la Colonisation, des Mines et des Pêcheries* (*RAMCMP*), 1917–18, 162; *RAMCMP,* 1919–20, 210.
27 Ibid., 1915–16, 130.
28 Kinsey, "Fashioning a Freshwater Eden," 267.
29 Archives de l'Université de Montréal, Fonds Vianney Legendre, P0271/series A4/file 8, "Rapport au conseil d'orientation économique au sujet de la pêche sportive dans la province," by L.A. Richard, deputy minister [1944]. See also Chartrand, Duchesne, and Gingras, *Histoire des sciences au Québec,* 290; Hébert, *Une histoire de l'écologie au Québec,* 418–19.

30 Côté, "Domestiquer le sauvage."
31 Harris, *A History of Higher Education in Canada*.
32 Hamelin, *Histoire de l'Université Laval*, 171. Robert Rumilly notes, however, that in his Speech from the Throne of 1882, Adolphe Chapleau announced the future creation of a mining school. This announcement did not lead to concrete action. Robert Rumilly, *Histoire de la province de Québec*, vol. 3, *Chapleau* (Montreal: Éditions Bernard Valiquette, 1940), 130n1.
33 On the sending of students to Ontario, see Émile Benoist, *L'Abitibi, pays de l'or* (Montreal: Éditions du Zodiaque, 1938), 152. On joining the Geological Survey of Canada, see Archives de l'École polytechnique de Montréal (AÉPM), 329-300-11/10, file "Dulieux, Émile," letter from É. Dulieux to the principal of the École polytechnique de Montréal, November 11, 1912.
34 Hodgetts, *Pioneer Public Service*, 229–35.
35 Heaman, *The Inglorious Arts of Peace*, 79–101; Zeller, *Inventing Canada*, 217–18.
36 Martin, *La chasse au Québec*; Ingram, *Wildlife, Conservation, and Conflict in Quebec*; Kinsey, "Fashioning a Freshwater Eden."
37 S. Castonguay, "Fédéralisme et centralisation de la recherche agricole au Canada."
38 The granting of university degrees at these early schools, however, was done by another institution, such as the University of Toronto or McGill University. Harris, *A History of Higher Education in Canada*.
39 Hodgetts, *From Arm's Length to Hands On*, 92.
40 Normand Perron provides an exhaustive portrait of the dissemination activities of the Department of Agriculture in *L'État et le changement agricole dans Charlevoix*, 89–116.
41 On Barnard's actions and the operation of the farmers' circles, see M.-A. Perron, *Un grand éducateur agricole*, 142–52. See also Bruno Jean, "Barnard, Edouard-André," in *Dictionary of Canadian Biography*, vol. 12 (Quebec City/Toronto: Université Laval/University of Toronto, 2003), http://www.biographi.ca/en/bio/barnard_edouard_andre_12E.html.
42 Hamel, Morisset, and Tondreau, *De la terre à l'école*.
43 *RAMA*, 1918–19, 137.
44 René Castonguay, "Perron, Joseph-Léonide," in *Dictionary of Canadian Biography*, vol. 15 (Quebec City/Toronto: Université Laval/University of Toronto, 2003), http://www.biographi.ca/en/bio/perron_joseph_leonide_15E.html.
45 Guy Boisclair provides an in-depth analysis of this reform. See his "Étude d'un mouvement de modernisation de l'agriculture."
46 *RAMA*, 1937–38, 22 (our translation).
47 Ibid., 35.
48 Provencher, *La Station de recherche de Deschambault*.
49 *RAMTF*, 1922–23, 32–33; Charland, *Les pâtes et papiers au Québec*.
50 *Statuts de la province de Québec*, 11 Geo. V (1921), c. 43, "Loi amendant les Statuts refondus, 1909, concernant le département de la Colonisation, des Mines et Pêcheries"; *RAMTF*, 1919–20, 33–34.
51 According to the Archives du Séminaire de Québec, Université 314/114, *Liste chronologique des diplômés de la Faculté des sciences et de la Faculté d'arpentage et de génie forestier*.
52 *RAMTF*, 1922–23, 26.
53 According to Avila Bédard, "Management of Crown Timberlands in Quebec," *Journal of Forestry* 36, 1 (January 1938): 934.
54 S. Castonguay, "Foresterie scientifique et reforestation."
55 Archives conserved at the Ministère des Ressources naturelles du Québec, Pépinière provinciale de Berthierville, "Rapports annuels de la pépinière de Berthierville" (1930–40). See also *RAMTF*, 1931–32, 184; *RAMTF*, 1937–38, 2.
56 *RAMTF*, 1929–30, 45.

57 Ibid., 62 (our translation).
58 Lionel Daviault, "L'organisation et les travaux du Laboratoire d'entomologie forestière de Berthierville," *La forêt québécoise* 4 (October 1942): 389–406.
59 Auger, "La recherche utilitaire dans les facultés de génie canadiennes," 25.
60 AÉPM, 999-305-02/60, dossier "M. Mailhiot, Adhémar," "Le laboratoire d'analyses du gouvernement" [n.d.].
61 John A. Dresser, "The Division of Geology, Quebec Bureau of Mines," *Canadian Institute of Mining and Metallurgy: Transactions* 35 (1932): 235.
62 RAMCMP, 1929-30, 514.
63 S. Castonguay, "Naturalizing Federalism."
64 Pross, "Development of Professions in the Public Service"; Rossiter, "Producing Provincial Space"; Brownstein, "Spasmodic Research as Executive Duties Permit."
65 Hodgetts, *From Arm's Length to Hands On*, 22.
66 Linteau, Durocher, and Robert, *Quebec*, 221–23, 478–81.
67 Nelles, *The Politics of Development*, 2; Hodgetts, *From Arm's Length to Hands On*, 70.
68 Gillis and Roach, *Lost Initiatives*, 180–82.
69 Kuhlberg, *One Hundred Rings and Counting*; Lambert and Pross, *Renewing Nature's Wealth*, passim.
70 For mines and forests, see Anholt, "An Administrative History of the British Columbia Government Agents"; Gillis and Roach, *Lost Initiatives*; Nelles, *The Politics of Development*; Rossiter, "Producing Provincial Space." For wildlife in Ontario and British Columbia, see, respectively, Knight, "Samuel Wilmot, Fish Culture, and Recreational Fisheries in Late 19th Century Ontario," and Peyton, "Imbricated Geographies of Conservation and Consumption in the Stikine Plateau."
71 On the virtues of these commissions of inquiry, see Hodgetts, *From Arm's Length to Hands On*, 9.

CHAPTER 2: THE INVENTION OF A MINING SPACE

1 On the beginnings and history of the Geological Survey of Canada, see Zaslow, *Reading the Rocks*.
2 Zeller, *Inventing Canada*.
3 Grek-Martin, "Survey Science on Trial."
4 Courville, *Immigration, colonisation et propagande*, 554.
5 Owram, *Promise of Eden*.
6 For a history of the mining industry in Quebec, see Vallières, *Des mines et des hommes;* Paquette, *Les mines du Québec*.
7 Zaslow, *The Opening of the Canadian North*.
8 Stafford, "Geological Surveys, Mineral Discoveries, and British Expansion"; Dupree, *Science in the Federal Government*, 91–114.
9 *Report of the Select Committee Appointed by the House of Commons to Obtain Information as to Geological Surveys, &c.* (Ottawa: Maclean, Roger, 1884).
10 For a biography of Dawson, see Suzanne Zeller and Gale Avrith-Wakeam, "Dawson, George Mercer," in *Dictionary of Canadian Biography*, vol. 13 (Quebec City/Toronto: Université Laval/University of Toronto, 2003), http://www.biographi.ca/en/bio/dawson_george_mercer_13E.html. Dawson's presentation is found in George Dawson, "On Some of the Larger Unexplored Regions of Canada," *Ottawa Naturalist* 4 (1890): 29–40.
11 Dawson, "On Some of the Larger Unexplored Regions," 29.
12 Ibid.

13 J.-C.-K. Laflamme, "Comptes rendus des observations géologiques faites dans la région du Saguenay" and "Compte rendu sommaire d'explorations dans la région du lac Saint-Jean," *Rapport des opérations: Commission géologique du Canada, Partie D, 1882–1884*, 1885, 1–8, 12–13; J.-C.-K. Laflamme, "Compte rendu sommaire d'explorations dans les comtés de Charlevoix et de Montmorency," *Rapport annuel: Commission géologique du Canada, Partie A, 1892–1893*, new series, 1893, 51–52; J.A. Dresser, "Preliminary Report on Shefford Mountain, Quebec," *Canadian Geological Survey: Summary Report, 1898, Part A*, 1899, 120–21; J.A. Dresser, "An Investigation of the Copper-Bearing Rocks of the Eastern Townships of the Province of Quebec," *Canadian Geological Survey: Summary Report, 1902, Part A*, 1903, 304318; J.A. Dresser, "The Serpentine Belt of the Eastern Townships, Que.," *Canadian Geological Survey: Summary Report, 1907*, 1908, 72–73; J.A. Dresser, "Serpentine Belt of Southern Quebec," *Canadian Geological Survey: Summary Report, 1909*, 1910, 180–200. For a biography of Laflamme, see Raymond Duchesne, "Laflamme, Joseph-Clovis-Kemner," in *Dictionary of Canadian Biography*, vol. 13 (Quebec City/Toronto: Université Laval/University of Toronto, 2003), http://www.biographi.ca/en/bio/laflamme_joseph_clovis_kemner_13E.html.
14 The Acton mine was apparently for some time the largest productive copper mine in the world. It was successfully operated for twelve years and then was closed. James Douglas, "Early Copper Mining in the Province of Quebec," *Canadian Mining Institute Journal* 13 (1910): 254–72, 270.
15 Zaslow, *Reading the Rocks*, 250–53.
16 In this chapter, I use the expression "Bureau of Mines" even though it was replaced by "Service of Mines" in the mid-1920s.
17 Nelles, *The Politics of Development*, 122–26.
18 Sutherland Brown, "A Century of Service," 32–33; Anholt, "An Administrative History of the British Columbia Government Agents," 104, 108.
19 Zaslow, *The Opening of the Canadian North*, 159.
20 *Statuts de la province de Québec*, 43–44 Vict. (1880), c. 12, art. 156. In Zaslow's view, the phosphate industry in the Ottawa Valley was likely responsible for adoption of the statute, whereas Jean Hamelin and Yves Roby see it more as the consequence of "frictions between gold seekers and owners." See Hamelin and Roby, *Histoire économique du Québec*, 257 (our translation).
21 Vallières, *Des mines et des hommes*, 55 (our translation).
22 Ibid., 51 (our translation).
23 *Résumé du discours prononcé par l'honorable M. Flynn lors de la deuxième lecture du projet de loi concernant les mines* ([Quebec], 1879), 4.
24 Drolet, "La Loi des mines de Québec"; Paquette, "Industries et politiques minières au Québec."
25 This provision would be extended to the entire territory. Lacasse, "La propriété des mines en droit québécois."
26 Marc Vallières, "Obalski, Joseph," in *Dictionary of Canadian Biography*, vol. 14 (Quebec City/Toronto: Université Laval/University of Toronto, 2003), http://www.biographi.ca/en/bio/obalski_joseph_14E.html.
27 Grek-Martin, "Survey Science on Trial," 4.
28 In his 1880 annual report, the inspector of gold mines, H.J.J. Duchesnay, had recommended the hiring of a mineralogist to "make an authentic assay of the quartz samples sent to him." *RACTC*, 1880, 38 (our translation).
29 Thépot, "Les ingénieurs du corps des mines."
30 Braun, "Producing Vertical Territory."
31 He drew his first report from it after he was hired by the commissioner of Crown lands: "Appendice no 47: Notes sur les produits minéraux de la province de Québec," in *RACTC*, 1882.

32 Titled *Rapport sur les mines de la province de Québec,* this publication became *Opérations minières dans la province de Québec* in 1901. BANQ-Québec, Fonds ministère de l'Énergie et des Ressources (E20), correspondances générales du Bureau des mines, acq. 1960-01037/285, vol. 292, rapport de compagnies minières pour 1892, 1893, 1895.
33 See the commentary by Jules Côté in *Rapport du secrétaire du "Bureau des mines," RACTC,* 1893–94, 75.
34 BANQ-Québec, E20, acq. 1960-01-037/389, J.-C.-K. Laflamme to J. Obalski, "Les puits à gaz à Louiseville," October 25, 1880.
35 E. Dulieux, "Rapport d'une exploration dans la région des lacs Chibougamau, Doré, David et Asichinibastat," *Opérations minières dans la province de Québec,* 1908, 52–87; E. Dulieux, "The Chibogomou [sic] Region, Province of Quebec," *Canadian Mining Institute, Transactions* 12 (1909): 184–93.
36 J.H. Valiquette, "Rapport du voyage d'exploration à la montagne Brillante dans la péninsule du Labrador," *Opérations minières dans la province de Québec,* 1908, 32–51.
37 *Statuts de la province de Québec,* 43–44 Vict. (1880), c. 12.
38 A.P. Low, "The Mistassini Region," *Ottawa Naturalist* 4 (1890): 12.
39 Rousseau, "Bataille de sextants autour du lac Mistassini."
40 On how these borders were determined, see Immarigeon, "Les frontières du Québec," 13–16.
41 Henry O'Sullivan, *Report of Progress of Exploration in the Country between Lake St. John and James Bay* (Quebec City: Charles Pageau, 1898), 3.
42 Morissonneau, *La Société de géographie de Québec,* 71–75, 92.
43 O'Sullivan, *Report of Progress,* 15. See also Morissonneau, *La Société de géographie de Québec.*
44 O'Sullivan, *Report of Progress,* 15.
45 Henry O'Sullivan, *Second Report of Progress of Exploration in the Country between Lake St. John and James Bay: Including the Region of Lake Mistassini and the Basins of the Great Nottaway and Rupert Rivers: Together with a Key Plan to Accompany Remarks on the Different Proposed Railways between Quebec and James Bay: Made under Instructions from the Department of Colonization and Mines, Quebec* (Quebec City: Charles Pageau, 1901).
46 *Rapport annuel du ministère des Terres, des Mines et des Pêcheries* (*RAMTMP*), 1905–06, xvii (our translation).
47 *RAMTMP,* 1904–05, 175 (our translation). Strangely, many describe this fieldwork as Obalski's first official exploration after more than twenty years working as an inspector. See Zaslow, *Reading the Rocks,* 231; Vallières, *Des mines et des hommes,* 67.
48 J. Obalski, "Rapport sommaire sur les opérations minières dans la province de Québec pour l'année 1904, conformément à l'article 1581 de la loi des mines de Québec," in *RAMTMP,* 1904, 175–76 (our translation).
49 J. Obalski, "District minier de Chibogomo [sic]," *Opérations minières dans la province de Québec,* 1905, 23–37; "Exploration au nord du comté de Pontiac," *Opérations minières dans la province de Québec,* 1906, 5–31.
50 Dulieux, "Rapport d'une exploration," 83 (our translation).
51 A.P. Low, *Geological Report on the Chibougamau Mining Region in the Northern Part of the Province of Quebec, 1905* (Ottawa: Government Printing Bureau, 1906), 5.
52 Among others, Morley E. Wilson, *Geology of an Area Adjoining the East Side of Lake Timiskaming Quebec* (Ottawa: Government Printing Bureau, 1910).
53 Alfred E. Barlow, *Second Edition of a Report on the Geology and Natural Resources of the Area Included by the Nipissing and Timiskaming Map-Sheets: Comprising Portions of the District of Nipissing, Ontario and of the County of Pontiac, Quebec* (Ottawa: Geological Survey of Canada, 1907); A.E. Barlow, "On the Quebec Side of Lake Temiskaming," *Canadian Geological Survey Summary Report,* 1906, 113–18.

54 Obalski, "District minier de Chibogomo," 23.
55 On Obalski's title, see BANQ-Québec, E20, 1960-01-037/298, Correspondances générales, Bureau des mines, dossier 392/06, "Obalski nommé surintendant des mines et chef du bureau des Mines de la Province en date du 17 janvier 1906."
56 Alfred Pellan, *Le Nouveau Québec: région du Témiscamingue: ressources agricoles, forestières, minières et sportives* (Quebec City: Dussault et Proulx, 1906); *Le Témiscamingue (Nouveau-Québec): ses ressources, ses progrès et son avenir* (Quebec City, 1910).
57 *The Province of Quebec: Its Products and Its Resources* (Quebec City: Department of Agriculture, 1911), 12.
58 J. Obalski, *Province de Québec: industries minérales, préparé spécialement pour l'Exposition de Liège, Belgique* (Quebec City: Gouvernement du Québec, 1905), 44 (our translation; emphasis added).
59 J. Obalski, *Province de Québec: industries minérales*, 11–12. See also J. Obalski, "A New Mining District in the North of Quebec," *Canadian Mining Institute, Transactions* 10 (1907): 363–66; "On the Probability of Finding Mines in Northern Quebec," *Canadian Mining Institute, Transactions* 9 (1906): 218–20.
60 Eugène Rouillard, "La région de l'Abitibi et de Chibougamo," *Bulletin de la Société de géographie du Québec*, New Series 3, 1 (January-February 1908): 38 (our translation). See also BANQ-Québec, E20, 1960-01-037/298155/06, H. Mortimer-Lamb to Obalski, January 9, 1906.
61 J. Obalski, *Province de Québec: industries minérales*, 44. See also J. Obalski, "New Discoveries in Northern Quebec," *Canadian Mining Institute, Transactions* 10 (1907): 105–7.
62 A.P. Low, "Report on the Chibougamau Mining Region in the Northern Part of the Province of Quebec," *Opérations minières dans la province de Québec*, 1905, 24–36.
63 Ibid., 61.
64 A.E. Barlow, *Rapport préliminaire de la Commission d'études sur la région du lac Chibougamou* [sic] (Quebec City: Ministère de la Colonisation, des Mines et des Pêcheries, 1911).
65 *Report on the Geology and Mineral Resources of the Chibougamau Region, Quebec by the Chibougamau Mining Commission* (Quebec City: Government Printing Bureau, 1911), 13, 51.
66 Ibid., 13 and 14.
67 *Débats de l'Assemblée législative (débats reconstitués)*, 12th Legislature, 2nd Session (March 15, 1910 – June 4, 1910), Friday, April 8, 1910 (our translation).
68 *Report on the Geology and Mineral Resources of the Chibougamau Region*, 14.
69 Ibid., 71.
70 Ibid., 64–69.
71 *Débats de l'Assemblée législative (débats reconstitués)*, 12th Legislature, 4th Session (January 9, 1912 – April 3, 1912), Monday, March 4, 1912 (our translation).
72 J.-T. Larochelle, *La région minière de l'ouest de Québec* (Quebec City: Service des mines, 1938), 3.
73 On the opening up of the territory, see Boudreau, Courville, and Séguin, *Le territoire*, 77.
74 Brunelle, *Histoire de l'administration québécoise*, 37 (our translation).
75 P.É. Dulieux, "Rapport sur les dépôts de fer sur la Côte nord du St-Laurent," *Opérations minières dans la province de Québec*, 1911, 81–148; *Les minerais de fer de la province de Québec: gisements et utilisation* (Quebec City: E.E. Cinq-Mars, 1915).
76 Adhémar Mailhiot, "Rapport géologique sur une partie de la région de Gaspé, bassins des rivières York et Ste Anne 1910," *Opérations minières dans la province de Québec*, 1910, 91–99.
77 R. Harvie, "Géologie et minéralogie du district du lac Opasatica," *Opérations minières dans la province de Québec*, 1910, 82–91.
78 J.A. Bancroft, "Rapport sur la géologie et les ressources minérales de la région des lacs Keekeek et Kewagama," *Opérations minières dans la province de Québec*, 1911, 175–230.
79 *Rapport annuel du ministère de la Colonisation et des Mines (RAMCM)*, 1900–01, 238 (our translation).

80 "En passant," *Canadian Mining and Mechanical Review* 15, 9 (September 1896): 195.
81 AÉPM, 100-300-12, *Procès-verbaux de la Corporation de l'École polytechnique de Montréal*, December 5, 1910, 249–50; Auger, "La recherche utilitaire dans les facultés de génie canadiennes," 25, 121–22.
82 T.C. Denis, *Opérations minières dans la province de Québec*, 1909, 7–9; S. Dufault, *Guide du mineur* (Quebec City: Ministère de la Colonisation, des Mines et des Pêcheries, 1910), 32–34.
83 Bureau of Mines of Quebec, *Descriptive Report of the Gold Deposits of Lake Demontigny, Abitibi* (Quebec City: L.A. Proulx, 1922); H.C. Cooke, W.F. James, and J.B. Mawdsley, "Geology and Ore Deposits of Rouyn-Harricanaw Region, Quebec," *Memoir of the Geological Survey*, 166 (Ottawa: King's Printer, 1931).
84 Benoît-Beaudry Gourd, "L'Abitibi-Témiscamingue minier: 1910–1950," in Vincent et al., *L'Abitibi-Témiscamingue*, 288–304.
85 Vallières, "Les entreprises minières québécoises et la grande dépression des années 1930," 133–47.
86 C. Girard and Perron, *Histoire du Saguenay–Lac-Saint-Jean*, 420–21.
87 In fall 1928, Minister Perrault announced, at the annual convention of the Canadian Institute of Mining and Metallurgy, his intention to expand the Bureau of Mines by adding a geology division. John A. Dresser, "The Division of Geology, Quebec Bureau of Mines," *Canadian Institute of Mining and Metallurgy, Transactions* 35 (1932): 233–39.
88 *Rapport annuel du ministère des Mines* (RAMM), 1929–30, 16 (our translation). See also *Opérations minières dans la province de Québec*, 1928, 16–19.
89 I.W. Jones, "Progress of Geological Investigation in Quebec," *Canadian Mining Journal* 69, 10 (October 1948): 116–26, quoted in Brunelle, *Histoire de l'administration québécoise*, 56 (our translation).
90 Brunelle, *Histoire de l'administration québécoise*, 56 (our translation).
91 "Gold Mining Quebec's Best Colonizer," *Canadian Mining Journal* 44 (1923): 735; "More Technical Officers for Quebec Bureau of Mines," *Canadian Mining Journal* 44 (1923): 879–80; "Salaries in the Quebec Bureau of Mines," *Canadian Mining Journal* 45 (1924): 85–86; "Wake up Quebec," *Canadian Mining Journal* 45 (1924): 1186; "Proposed Quebec Mineral Resources Bureau," *Canadian Mining Journal* 46 (1925): 342.
92 After leaving politics in 1936, Perrault became a member of the executive committees of Noranda Mines, Hallnor Mines, Pamour Porcupine Gold Mines, Amulet Dufault Mines, Waite Amulet Mines, and Noranda Copper and Brass. Deschênes, *Dictionnaire des parlementaires du Québec*, 595. After the fall of Alexandre Taschereau's Liberal government in 1936, the mines dossier was associated with the fisheries dossier within the provincial administration until 1939.
93 "Quebec Requires a Department of Mines," *Canadian Mining Journal* 50 (1929): 855; "A Quebec Department of Mines," *Canadian Mining Journal* 50 (1929): 1129; "Status of Quebec Bureau of Mines Is Much Improved," *Canadian Mining Journal* 51 (1930): 359.
94 Dresser, "The Division of Geology, Quebec Bureau of Mines," 235.
95 Brunelle, *Histoire de l'administration québécoise*, 67 (our translation).
96 Ibid., 57 and 67; I.W. Jones, "Progress of Geological Investigation in Quebec," *Canadian Mining Journal* 69, 10 (October 1948): 122.
97 A.-O. Dufresne, "L'industrie minérale de Québec," *Bulletin de la Société de géographie du Québec* 23 (1929): 24 (our translation).
98 Brunelle, *Histoire de l'administration québécoise*, 61.
99 McGill University Archives, archives of Professor T.H. Clark MG-3055, acc. 93-048, vol. 1, dossier 37, letter from J.A. Dresser to T.H. Clark, May 25, 1938; dossier 16, letter from I.W. Jones to T.H. Clark, April 6, 1939.
100 Dresser, "The Division of Geology, Quebec Bureau of Mines," 235.

101 W.E. Hale, "Memorial of Graham Stewart MacKenzie," *American Mineralogist* 46 (1961): 501–4; Petroleum Industry Oral History Project, transcription of interview with Robert A. Brown by Aubrey Kerr, May 1982, 9.
102 T.-C. Denis, "B-Esquisse géologique," in "Description de la province," *Annuaire statistique de la province de Québec* (Quebec City: L.A. Proulx, 1914), 16; Théophile-C. Denis, *Esquisse géologique et minéraux utiles de la province de Québec* (Quebec City: Ministère de la Colonisation, des Mines et des Pêcheries, bureau des mines de la province de Québec, n.d.). Although the publication has no official date, the author refers several times to studies underway in 1924. The maps are reproduced in George Albert Young, *Geological Map of the Dominion of Canada and Newfoundland*, Publication no. 1277, 3rd ed. (Ottawa: Department of Mines of Canada, Geological Survey of Canada, 1924).
103 *RAMM*, 1932–33, xxii–xxiii.

CHAPTER 3: SOIL CLASSIFICATION AND SEPARATION OF FOREST AND COLONIZATION AREAS

1 E.T. Fletcher, "Crown Land Survey, P.Q.," *Bulletin de la Société de géographie du Québec* 1 (1885): 42–43.
2 Eugène Rouillard, "La région de l'Abitibi et de Chibougamo," *Bulletin de la Société de géographie du Québec* N.S. 3, 1 (January-February 1908): 25 (our translation).
3 On the discourses around depletion of the forest, see Olson, *The Depletion Myth*.
4 Hays, *Conservation and the Gospel of Efficiency*; Balogh, "Scientific Forestry and the Roots of the Modern American State"; Pisani, "The Many Faces of Conservation."
5 Hardy, "Exploitation forestière et environnement au Québec"; M. Girard, "La forêt dénaturée."
6 Gaudreau, "Exploitation des forêts publiques au Québec," 9.
7 *RAMTF*, 1904–05, 128.
8 *Proceedings of the American Forestry Congress at Its Sessions Held at Cincinnati, Ohio, in April 1882 and at Montreal, Canada, in August 1882* (Washington, DC, printed for the Society, 1883). On the organization of the congress and its effects on forestry policy in Canada, see Gillis and Roach, *Lost Initiatives*, 31–49. Although a number of historians see the fact that this meeting was held in Montreal as an indication that American conservationism was making inroads in Canada, the site was chosen solely because the annual congress of the American Association for the Advancement of Science, with which the American Forestry Congress was associated, was taking place in the city.
9 *Statuts de la province de Québec*, 46 Vict. (1883), c. 9 and c. 10, "Acte pour amender de nouveau le chapitre 23 des statuts refondus du Canada concernant la vente et l'administration des bois croissant sur les terres publiques" and "Acte pour pourvoir d'une manière plus efficace aux moyens de prévenir les feux de forêts." According to Luc Bouthillier, the reserves originated in the incorporating statute for lands and forests, *Statuts de la province de Québec*, 30 Vict. (1875), c. 11. See Bouthillier's PhD dissertation, "Le concept de rendement soutenu en foresterie dans un contexte nord-américain."
10 *Statuts de la province de Québec*, 51–52 Vict. (1895), c. 15, "Acte relatif à la vente et l'administration des terres publiques, aux bois et aux mines, ainsi qu'au défrichement des terres et à la protection des forêts."
11 Hodgins, Benidickson, and Gillis, "The Ontario and Quebec Experiment in Forest Reserves," 22–23.
12 *Statuts de la province de Québec*, 58 Vict. (1895), c. 22 and c. 23.
13 W.C.J. Hall, "The Laurentide National Park," *Canadian Forestry Association: Report of the Annual Meeting, 1901*, 81.

14 Ibid.
15 Hardy and Séguin, *Forêt et société en Mauricie*.
16 On the forestry policies of the Ontario government at the time, see Nelles, *The Politics of Development*, 182–214.
17 Gow, *Histoire de l'administration publique québécoise*, 94.
18 On the Colonization Commission and on soil classification for forestry and agricultural colonization, see Flamand-Hubert, "La forêt québécoise en discours dans la première moitié du XXe siècle," 92–113.
19 *Rapport de la Commission de la colonisation de la province de Québec* (Quebec City: Charles Pageau, 1904), 77–78 (our translation).
20 Henry Joly de Lotbinière, "The Danger Threatening the Crown Lands Forests of the Province of Quebec through the Cutting of Pulp Wood as at Present Sanctioned by the Regulations Concerning Wood and Forests," *Canadian Forestry Association: Report of the Annual Meeting, 1906*, 107; Jean Chrysostome Langelier, "The Pulp and Paper Industry in Relation to Our Forests," *Canadian Forestry Association: Report of the Annual Meeting, 1906*, 39–47; J.-C. Chapais, "La forêt et le cultivateur, conférence donnée devant la société pomologique de la province de Québec, au collège McDonald [sic] le 10 décembre 1909," Quebec City, 1910.
21 Jean Chrysostome Langelier, *Richesse forestière de la province de Québec* (Quebec City, 1905), 53–57.
22 *Statuts de la province de Québec*, 11 Geo. V (1921), c. 43, "Loi amendant les Statuts refondus, 1909, concernant le département de la Colonisation, des mines et pêcheries"; *RAMTF,* 1920–21, 33–34; *RAMTF,* 1923–24, 26.
23 *Statuts de la province de Québec*, 6 Ed. VII (1906), c. 15, s. 4, para. 1-a. *Des réserves de forêts et de la coupe du bois sur icelles*, "Loi amendant la loi concernant les terres publiques."
24 *L'administration libérale: discours prononcé par l'hon. M. Lomer Gouin, ministre de la colonisation et des travaux publics, à l'Assemblée législative de Québec le 24 mars 1904* (Quebec City, 1904), 83 (our translation).
25 *Statuts de la province de Québec*, 6 Ed. VII (1906), c. 17, "Loi établissant une réserve de forêt, de chasse et de pêche dans la Gaspésie." See also "The Forest Reserves of the Province of Quebec," *Canadian Forestry Journal* 3, 2 (June 1907): 68.
26 Department of Agriculture of Quebec, *The Province of Quebec: Its Products and Its Resources* (Quebec City: Telegraph Printing, 1911), 9.
27 Demeritt, "Scientific Forest Conservation and the Statistical Picturing of Nature's Limits."
28 Hall, "The Laurentide National Park," 81. On the reserves, see also Martin, *La chasse au Québec*, 144; Hébert, "Conservation, culture et identité."
29 W.J.C. Hall, "The Gaspesian Forest Reserve," *Canadian Forestry Journal* 1, 3 (July 1905): 112, 110.
30 *RAMTMP,* 1904–05, 100. On hydroelectric developments during this period, see Bellavance, "L'État, la 'houille blanche' et le grand capital."
31 "The Forest Reserve of Quebec," *Canadian Forestry Journal* 3, 2 (July 1907): 69.
32 *Statuts de la province de Québec*, Geo. V (2nd session) (1911), c. 17, "Loi amendant les Statuts refondus, 1909, relativement à la création des réserves forestières cantonales." In the first annual report of the Forest Service, Piché pronounced himself in favour of this idea. *RAMTF,* 1909–10, 68.
33 *Rapport de la Commission de la colonisation de la province de Québec,* 85 (our translation).
34 Ibid. (our translation).
35 *RAMTF,* 1926–27, 31 (our translation).
36 For example, J.U. [sic] K. Laflamme, "Forestry Education," *Canadian Forestry Journal* 4, 1 (March 1908): 37.
37 Johnstone, *Forêts et tourments*, 17–29.

Notes to pages 79–81

38 A third, private school was established on the Biltmore Estate in North Carolina, owned by George Vanderbilt, a wealthy Pennsylvanian industrialist. James G. Lewis, "Trained by Americans in American Ways: The Establishment of Forestry Education in the United States, 1885–1921" (PhD dissertation, Florida State University, 2001).
39 Kuhlberg, *One Hundred Rings and Counting;* Glen Jordan and Graham R. Powell, "One Hundred Years of Forestry Education at UNB (1908–2008)," *Forestry Chronicle* 84 (2008): 478–80.
40 B.E. Fernow, *Lectures on Forestry* (Kingston: British Whig, 1903). On Fernow, see Rodgers, *Bernhard Eduard Fernow.*
41 The translation was published as *La forêt: conférences par M. le professeur B.E. Fernow, L.L.D., à l'École de minéralogie de Kingston, Ont., 26-30 janvier 1903* ([Quebec City]: Département des Terres et Forêts de la province de Québec, 1906). On the attempt at Queen's University to found a forestry school and its eventual failure in the face of the University of Toronto's efforts, see Kuhlberg, *One Hundred Rings and Counting,* 12–19.
42 Laflamme, "Forestry Education," 31–38.
43 ASQ, Université/61/111, "Conférence faite à l'Université Laval, 18 mars 1908" (our translation).
44 At least, this is the subject of a letter that he wrote to the chief forester of the United States, the dean of the forestry school at Yale, and the director of the École nationale des eaux et forêts de Nancy. ASQ, Université/68/15, J.-C.-K. Laflamme to Guifford Pinchot, [n.d.]; Chs.-O. Guyot to J.-C.-K. Laflamme, July 29, 1904; Université/62/120, Lomer Gouin to J.-C.-K. Laflamme, July 19, 1905; ASQ, Université/62/121, Henry Solon Graves to J.-C.-K. Laflamme, June 6, 1905.
45 See BANQ-Québec, E21, S 10, versement 1960 0108/515, lettre 26853; versement 1960 0108/522, lettre 29942; versement 1960-01-08/529, lettre 7008 and lettre 15239. At the forestry school itself, there was a course on forest seeding and plantation. The professor of the course, James William Toumey, was the author of *Seeding and Planting: A Manual for the Guidance of Forestry Students, Foresters, Nurserymen, Forest Owners, and Farmers* (New York: John Wiley, 1916), which was republished in 1931.
46 *RAMTF,* 1907–08, 71 (our translation).
47 BANQ-Québec, E21, S 10, versement 1960-01-038/529, lettre 19390, Avila Bédard and G.C. Piché to Adélard Turgeon, New Haven, October 16, 1906, reproduced in *RAMTF,* 1905–06, 63–65.
48 Reproduced in *RAMTF,* 1905–06, 69–72 (our translation).
49 Ibid., 72 (our translation).
50 *RAMTF,* 1906–07, 73 (our translation).
51 Letter from M.G.-C. Piché, forest engineer, regarding the establishment of a nursery for forest trees, Montreal, October 3, 1907, appendix 25, *RAMTF,* 1907–08, 68–69 (our translation).
52 *RAMTF,* 1907–08, 63 (our translation).
53 Mgr J.U.K. [sic] Laflamme, "Forestry Education," *Canadian Forestry Association: Report of the Annual Meeting, 1906,* 164, 160–71; *RAMTF,* 1907–08, 71.
54 Laflamme, "Forestry Education," trans. J.W. Michaud, "Un précurseur: Mgr Laflamme recteur de l'Université Laval," *Forêt conservation* 26, 10 (1960): 11 (our translation).
55 ASQ, Université/61/107, rapport de M.G.-C. Piché à l'hon. Allard, May 14, 1909 (our translation). See also G.-C. Piché, "Training Forest Students," *Canadian Forestry Journal* 4, 4 (December 1908): 186–89.
56 ASQ, Université/61/109, Piché to Laflamme, January 6, 1909; Université/61/108, Piché to Laflamme, February 12, 1909; Université/61/104, Piché to Laflamme, February 28, 1909; Université/61/105, Piché to Laflamme, March 3, 1909.
57 ASQ, Université/61/104, Piché to Laflamme, February 28, 1909 (our translation). And Piché added, "Our school will be 'voted,' so watch out for them" (our translation).

58 In 1918, Bédard replaced Piché, who wanted to devote himself fully to the Forest Service.
59 BANQ-Québec, E21, S 10, versement 1960/01-038/559 letter 1424/10, J.-C. Langelier to minister of lands and forests, December 5, 1909 (our translation). See also letter 4052/09, J.-C. Langelier to minister of lands and forests, March 7, 1909.
60 BANQ-Québec, E21, S 10, versement 1960/01-038/559, letter 18884/10, G.-C. Piché to minister, June 3, 1910 (our translation).
61 BANQ-Québec, E21, S 10, versement 1960/01-038/559, letter 18884/10, Paul Blouin to the minister of lands and forests, June 3, 1910. See also *RAMTF,* 1909–10, viii.
62 "Laval Forest School," *Canadian Forestry Journal* 10, 7 (August-September 1914): 93.
63 According to ASQ, Université 314/114, Liste chronologique des diplômés de la Faculté des sciences et de la faculté d'Arpentage et de génie forestier. *Statuts de la province de Québec,* 14 Geo. V (1924), c. 27, s. 3.
64 ASQ, Université/61/106, Piché to Allard, [n.d.] [spring 1909] (our translation).
65 Ibid. (our translation).
66 On the "puerile discussions," see Gaudreau, "L'État, le mesurage du bois et la promotion de l'industrie papetière."
67 ASQ, 9/223/9, Procès-verbaux du Conseil de l'Université Laval, excerpt from the meeting of February 27, 1914 (our translation).
68 *RAMTF,* 1910–11, 54; "Quebec Provincial Nurseries: Description of the Forest Nursery Station at Berthierville," *Canadian Forestry Journal* 9, 10 (October 1913): 149; *RAMTF,* 1909–10, 52; *RAMTF,* 1922–23, 32; *RAMTF,* 1931–32, 63.
69 Avila Bédard, "La pépinière de Berthierville," *Bulletin de la Société de géographie du Québec* 5 (1911): 117–18.
70 *RAMTF,* 1909–10, 54; "Quebec's Planting Operations: Waste Lands near Lachute Being Reforested," *Canadian Forestry Journal* 9, 7 (July 1913): 108.
71 Avila Bédard, "Les dunes de Lachute," *Bulletin de la Société de géographie du Québec* 5 (1911): 20–23; "Putting Useless Land to Work: How the Quebec Government Is Planting Up the Lachute Sand Plains. A Visit to Berthierville," *Canadian Forestry Journal* 11, 8 (August 1915): 147–50.
72 *RAMTF,* 1909–10, 66 (our translation).
73 Ibid., 65 (our translation).
74 G.C. Piché, "Forest Planting in Quebec," in *Forest Protection in Canada, 1912,* ed. Commission of Conservation of Canada (Ottawa: Mortimer, 1912), 135; *RAMTF,* 1916–17, 38; G.-C. Piché, "Les forêts du Canada," *Revue trimestrielle canadienne* 3 (1917): 204.
75 For instance, in Quebec, Laurentide in Grand-Mère, Riordon Pulp and Paper in Saint-Jovite (International Paper), Brompton Pulp and Paper in Bromptonville, Singer Manufacturing in Thurso, Price Brothers and Brown in Lac-Beauport, Donaconna Paper in Donnacona, Pejepscot Paper in Cookshire, and St. Lawrence Paper and Wayagamack Pulp and Paper in Trois-Rivières. G.C. Piché, "Notes sur la pépinière de Berthierville," *La vie forestière et rurale* 2 (1923): 277–83; *RAMTF,* 1908–09, 71; *RAMTF,* 1915; *RAMTF,* 1915–16, 39; *RAMTF,* 1916–17, 38; *RAMTF,* 1925–26, 33; *RAMTF,* 1927–28, 43; "Private Initiative in Replanting: What the Pejepscot Paper Company Is Doing," *Canadian Forestry Journal* 9, 3 (March 1913): 37.
76 On Laurentide, see Hardy and Séguin, *Histoire de la Mauricie,* 698–99.
77 Archives conserved at the Ministère des Ressources naturelles du Québec, Pépinière Proulx (AMPP), Ellwood Wilson to George Cahoon, Forest Planting Work of Laurentide Co. Ltd., January 9, 1919; Laurentide Company Ltd., Reforestation Policy [1923]. See also Ellwood Wilson, "A Forester's Work in a Northern Forest," *Forestry Quarterly* 7 (1909): 2–14.
78 BANQ-Québec, E21, S 10, versement 1960-01-038/559, G.-C. Piché to the minister of lands and forests, April 13, 1909; lettre 5991/09, Wilson to the minister of lands and forests, April 6, 1909.

79 AMPP, Rapport d'Ellwood Wilson, October 15, 1930; BANQ-Montréal, P149, versement 1983-03-011/478, dossier CPPC8.3, "correspondance, Ellwood Wilson," General discussion of raw material situation of Canada Power & Paper Corporation and Reforestation activities of the Laurentide Division [Ellwood Wilson report], September 3, 1931. For a history of the Laurentide nursery, see J.D. Gagnon, *Les plantations de Grand'Mère: Modèle de reboisement pour l'avenir* (Quebec City: Centre de recherches forestières des Laurentides, July 1972).

80 Ellwood Wilson, "Spruce Problems in Eastern Canada," *Forestry Chronicle* 3, 4 (December 1927): 23–26; R.W. Lyons, "Artificial Regeneration of White Spruce," *Forestry Chronicle* 1, 2 (December 1925): 9–19.

81 See, for example, G.-C. Piché, "Reboisement et plantations, 1ère partie: pourquoi faut-il reboiser et quels arbres doit-on employer dans les plantations," Département des Terres et Forêts, Service forestier, *Circulaire* 1 (1928).

82 Ellwood Wilson, "Reforestation in Canada: A Brief Resumé of Planting Activities throughout Canada: Responsibility for Next Forest Crop," *Illustrated Canadian Forest and Outdoors* 5, 11 (November 1929): 665–67.

83 AMPP, Ellwood Wilson to George Cahoon, Forest Planting Work of Laurentide Co. Ltd., January 9, 1919; Ellwood Wilson, "Planting Spruce for Commercial Purposes: Description of What the Laurentide Company Is Doing," *Canadian Forestry Journal* 11, 1 (January 1915): 3–6; Gagnon, *Les plantations de Grand'Mère;* Ellwood Wilson, "Reforestation in Canada," *Forestry Chronicle* 5, 2 (June 1929): 17.

84 *RAMTF*, 1931–32, 49. On the wave of consolidations in the paper mill industry, see Piédalue, "Les groupes financiers et la guerre du papier au Canada."

85 AMPP, Lettre à Maurice Moreau, ing. f., directeur du Service extérieur, MTF, le 23 décembre 1965, par Lucien Castonguay, ing. f., chef du district, concernant le projet de reconstruction à la pépinière de Grande-Piles; *RAMTF*, 1935–36, 51.

86 BANQ-Montréal, Fonds de la Consolidated-Bathurst, P149, versement 1983-03-011/478, dossier CPPC8.3, "correspondance, Elwood Wilson," George McKee, v.-p. Operations, to Ellwood Wilson, October 27, 1931.

87 B.F. Avery, "Ellwood Wilson 1872–1952," *Forestry Chronicle* 29, 1 (March 1953): 93–94.

88 *RAMTF*, 1907–08, 55–61.

89 *RAMTF*, 1918–19, 29–30.

90 *RAMTF*, 1924–25, 33.

91 *RAMTF*, 1919–20, 25 (our translation).

92 *Statuts de la province de Québec*, 15 Geo. V (1925), c. 32, "Loi modifiant les Statuts refondus, 1909," with respect to reforestation (our translation).

93 *RAMTF*, 1909–10, 141.

94 *RAMTF*, 1922–23, 29 (our translation).

95 Ibid. (our translation).

96 *RAMTF*, 1922–23, 29 (our translation). Piché envisaged such a system in 1913, but the war probably delayed its application. See "Forestry in Quebec," *Canadian Forestry Journal* 9, 11 (November 1913): 167–68.

97 J.-A. Roy, "Étude sur le reboisement dans la province de Québec," *Forestry Chronicle* 12, 3 (September 1936): 315–28.

98 Biays, "L'œkoumène agricole au Lac Saint-Jean," 105 (our translation).

99 Murdo MacPherson and Serge Courville, "Colonisation et coopération," in *Atlas historique du Canada*, vol. 3, *Jusqu'au cœur du XXe siècle, 1891–1961*, ed. Donald Kerr and Deryck W. Holdsworth (Montreal: Presses de l'Université de Montréal, 1990), pl. 44.

100 *RAMTF*, 1922–23, 27 (our translation). See also *RAMTF*, 1926–27, 26.

101 *RAMTF*, 1929–30, 35; C.-G. Piché, "La séparation du domaine forestier et du domaine agricole," *La forêt et la ferme* 2, 2 (February 1927): 51–52, 63–64.

102 C.-G. Piché, "Le reboisement dans la province de Québec," *La vie forestière* 5, 4 (June 1930): 121 (our translation).
103 *RAMTF*, 1931–32, 49; *RAMTF*, 1933–34, 64; *RAMTF*, 1938–39, 52–53.

Chapter 4: Surveillance and Improvement of Fish and Game Territories

1 *Rapport annuel du ministère des Terres, des Forêts et des Pêcheries* (*RAMTFP*), 1900–01, 82.
2 *Rapport annuel du commissaire des Terres de la Couronne* (*RACTC*), 1892–93, v (our translation).
3 *RAMTFP*, 1899–1900, 28. Starting in the 1920s, freshwater sportfishing was clearly distinguished from commercial fishing in official government designations. This chapter concerns only sportfishing.
4 Kinsey, "Fashioning a Freshwater Eden."
5 During the period under study, the department responsible for hunting and fishing evolved as follows: Commissioner of Lands, Forests, and Fisheries (1897–1901); Department of Lands, Mines, and Fisheries (1901–05); Department of Colonization, Mines, and Fisheries (1905–30); Department of Colonization, Hunting, and Fisheries (1930–35); Department of Public Works, Hunting, and Fisheries (1935–36); Department of Mines and Fisheries (1936–40).
6 Gagnon, *L'échiquier touristique Québécois*.
7 Ingram, *Wildlife, Conservation, and Conflict in Quebec*, 47–49.
8 S. Castonguay and Kinsey, "The Nature of the Liberal Order Framework," 233.
9 Kinsey, "Fashioning a Freshwater Eden," 203.
10 Aside from the works by Kinsey and Ingram cited above, see Martin, *La chasse au Québec*; Guay, *Histoires vraies de la chasse*; Gingras, *A Century of Sport*; Pascal Gagnon, "La pratique de la chasse dans le comté de Rimouski" (master's thesis, Université du Québec à Trois-Rivières, 2002).
11 *Statutes of Canada*, 22 Vict. (1858), c. 86, "Fishery Act"; Ingram, *Wildlife, Conservation, and Conflict*, 46.
12 Reigert, *American Sportsmen and the Origins of Conservation*; Dunlap, *Saving America's Wildlife*.
13 Irene Bilas, "Fortin, Pierre-Étienne," in *Dictionary of Canadian Biography*, vol. 11 (Quebec City/Toronto: Université Laval/University of Toronto, 2003), http://www.biographi.ca/en/bio/fortin_pierre_etienne_11E.html.
14 *Rapport annuel du ministère de la Marine et des Pêcheries du Canada* (*RAMMPC*), 1874, lxxiii.
15 Geneviève Massicotte, "Rivalités autour de la pêche au saumon sur la rivière Ristigouche: étude de la résistance des Mi'gmaq (1763–1858)" (master's thesis, Université du Québec à Montréal, 2009).
16 Peter Gossage, *Water in Canadian History: An Overview* ([Ottawa]: Inquiry on Federal Water Policy, 1985).
17 This was the case of *The Queen v Robertson* (1882) 6 SCR 52. See Kinsey, "Fashioning a Freshwater Eden," 142–45.
18 Jean Bouffard, *Traité du domaine* (Quebec City: Imprimerie Le Soleil, 1921), 159.
19 Lacasse, "Réserve des trois chaînes et gestion du domaine public foncier au Québec."
20 Côté, "Domestiquer le sauvage."
21 Martin, *La chasse au Québec*.
22 Ingram, *Wildlife, Conservation, and Conflict*, 112–17.
23 On the rights of settlers in these territories, see Débats de l'Assemblée législative (débats reconstitués), 8th Legislature, 4th Session (October 30, 1895–December 21, 1895), Tuesday, December 10, 449–51. On the statute respecting the Triton Club, see "Une loi pour amender la loi de chasse du Québec," *Statuts de la province de Québec*, 59 Vict. (1895), c. 20.

24 Ingram, *Wildlife, Conservation, and Conflict*, 214; Anonymous, *The Fish and Game Clubs of the Province of Quebec: What They Mean to the Province, What Privilege They Enjoy* (Quebec City: Département de la colonisation, des mines et des pêcheries du Québec, 1914).
25 Gagnon, "L'appropriation ludique de la forêt au Québec."
26 *Statuts de la province de Québec*, 48 Vict. (1885), c. 12.
27 Côté, "Domestiquer le sauvage."
28 Anonymous, *The Fish and Game Clubs*.
29 On the extinction of these species, see Pauly, *Biologists and the Promise of American Life*, 161–64; Isenberg, *The Destruction of the Bison*.
30 Ingram, *Wildlife, Conservation, and Conflict*, 96.
31 *RACTC*, 1884–85, 132 (our translation).
32 *RACTC*, 1885–86, 78 (our translation).
33 *RACTC*, 1894–95, ix (our translation).
34 *Rapport annuel du commissaire des Terres de la Couronne, des Mines et des Pêcheries* (*RACTCMP*), 1902–03, 36 (our translation). Joncas's mandate lasted from November 17, 1896, to March 28, 1903. Marc Desjardins, "Joncas, Louis-Zéphirin," in *Dictionary of Canadian Biography*, vol. 13 (Quebec City/Toronto: Université Laval/University of Toronto, 2003), http://www.biographi.ca/en/bio/joncas_louis_zephirin_13E.html.
35 *RACTCMP*, 1903–04, 36 (our translation).
36 Ibid., 31–34 (our translation).
37 *RAMCMP*, 1931–32, 131–32; *RAMCMP*, 1932–33, 184.
38 *RAMCP*, 1929–30, vi–viii (our translation).
39 Gérard Delorme, "Pêche et chasse sportives," in *Pêche et chasse – étude préparée avec la collaboration du Département des pêcheries maritimes et du Département de la chasse et de la pêche de Québec, ainsi que de l'École supérieure des pêcheries de Sainte-Anne-de-la-Pocatière*, ed. Esdras Minville (Montreal: Fides and École des hautes études commerciales, 1946), 336 (our translation).
40 Ingram, *Wildlife, Conservation, and Conflict*, 181; Kinsey, "Fashioning a Freshwater Eden," 250.
41 Ingram, "'Au temps et dans les quantités qui lui plaisent.'"
42 See, among others, the accounts by Napoléon Alexandre Comeau in *Life and Sport on the North Shore of the Lower St. Lawrence and Gulf, Containing Chapters on Salmon Fishing, Trapping, the Folk-Lore of the Montagnais Indians and Tales of Adventure on the Fringe of the Labrador Peninsula* (Quebec City: Daily Telegraph, 1909).
43 *RAMCMP*, 1911–12, 159 (our translation).
44 S. Castonguay and Kinsey, "The Nature of the Liberal Order Framework," 233.
45 E.-É. Taché, *Chasse et pêche: Province de Québec: temps de prohibition* (Quebec City: Imprimerie générale A. Côté, February 20, 1895), 4 (our translation; emphasis in original).
46 Other measures had a similar effect, including adoption of more restrictive regulations regarding marketing of the product of hunting and fishing (ban on selling caribou meat and trout).
47 S. Castonguay and Kinsey, "The Nature of the Liberal Order Framework," 234.
48 Ibid.
49 Ingram, *Wildlife, Conservation, and Conflict*, 181; Kinsey, "Fashioning a Freshwater Eden," 250.
50 Côté, "Domestiquer le sauvage."
51 BANQ-Québec, Fonds ministère des Terres et Forêts, E21, S74, SS5, Dossier 7683/10, letter from Dr. A. Bouillon to the Minister of Colonization, Mines, and Fisheries, July 15, 1910, regarding logs belonging to the Price Brothers Company that obstructed the Matane River and the destruction "by cinders falling into the river from chimneys that are too short and provided with smokeless incinerators" by the same company.

52 *Rapport annuel du commissaire des Terres, des Forêts et des Pêcheries (RACTFP)*, 1901–02, 37.
53 *RACTFP*, 1899–1900, 36.
54 Joncas, quoted in *RACTFP*, 1900–01, 39 (our translation); Caron quoted in *RAMCMP*, 1907–08, 152 (our translation).
55 *RAMCMP*, 1915–16, 133–34 (our translation).
56 North American Fish and Game Protective Association, *Minutes of the Proceedings of the First Convention Held in Montreal, 2nd February 1900* (Quebec City: Darveau et Beauchamp, 1906), 29; Shudam S. Hill, "To the Editor of Rod and Gun in Canada," *Rod and Gun Canada* 6, 10 (March 1905): 557–59, quoted in Ingram, *Wildlife, Conservation, and Conflict*, 246.
57 North American Fish and Game Protective Association, *Minutes of the Proceedings of the First Convention Held in Montreal*, 133.
58 Ibid., 138.
59 Order-in-Council, December 2, 1905; *RAMTF*, 1906, 100–1 (our translation).
60 *RAMTF*, 1904–05, 182 (our translation). On the proportion of parkland in timber berths, see Corporation professionnelle des ingénieurs forestiers du Québec, *Parcs, territoires et zones analogues: plein air et conservation au Québec* (Quebec City: Gouvernement du Québec, 1974).
61 *RAMTF*, 1906–07, 94 (our translation).
62 Ibid. (our translation).
63 In imperial figures, the park grew from 2,531 square miles to almost 3,700 square miles.
64 Gagnon, "L'appropriation ludique."
65 *RAMCMP*, 1923–24, 98.
66 *RAMCMP*, 1919–20, 172 (our translation).
67 *RAMCMP*, 1922–23, viii (our translation).
68 *RAMTF*, 1935–36, 135.
69 Ibid., 136 (our translation).
70 Louis-Arthur Richard, "Les parcs nationaux," *L'Actualité économique* 14 (1938): 59 (our translation).
71 Ibid. (our translation).
72 BANQ-Québec, Fonds ministère du Loisir, de la Chasse et de la Pêche, E22, acq. 1976-04001/1, correspondence of L.A. Richard and O. Gagnon, [n.d.]; Delorme, "Pêche et chasse sportives" (our translation).
73 *RAMTFP*, 1905–06, 130–31.
74 *RAMTF*, 1906–07, 93.
75 *RAMTF*, 1908–09, 104–5.
76 Kinsey, "'Seeding the Water as the Earth.'"
77 Richard Nettle, *The Salmon Fisheries of the St Lawrence and Its Tributaries* (Montreal: John Lovell, 1857).
78 Alan B. McCullough, "Samuel Wilmot," in *Dictionary of Canadian Biography*, vol. 12 (Quebec City/Toronto: Université Laval/University of Toronto, [1990]), http://www.biographi.ca/en/bio/wilmot_samuel_12E.html.
79 Knight, "Samuel Wilmot, Fish Culture, and Recreational Fisheries in Late 19th Century Ontario."
80 E.E. Prince, *The Progress of Fish Culture in Canada*, Special Report (Ottawa: Government Printing Bureau, 1905).
81 *RAMCMP*, 1916–17, 130.
82 Ibid.
83 Darin Kinsey, "Chambers, Edward Thomas Davies," in *Dictionary of Canadian Biography*, vol. 16 (Quebec City/Toronto: Université Laval/University of Toronto, 2003), http://www.biographi.ca/en/bio/chambers_edward_thomas_davies_16E.html.

84 E.T.D. Chambers, *Fish and Game in the Province of Quebec: Our Rivers and Lakes* (Quebec City: Proulx & Proulx, 1895), and *The Angler's Guide to Eastern Canada: Showing Where, How and When to Fish Our Salmon, Bass, Ouananiche and Trout* (Quebec City: Morning Chronicle Office, 1898). Chambers was also credited with the 1914 publication on the benefits to the province of private hunting and fishing clubs (see note 24).
85 Kinsey, "Fashioning a Freshwater Eden," 176.
86 *RAMCMP,* 1917–18, 163.
87 B.W. Taylor, "Fish Cultural Activities of the Province of Quebec," *Transactions of the American Fisheries Society* 62 (1932): 224–27.
88 Georges Préfontaine, "Le développement des connaissances scientifiques sur les pêcheries maritimes et intérieures de l'est du Canada," *Actualité économique* 21 (1946): 238.
89 Knight, "Samuel Wilmot, Fish Culture, and Recreational Fisheries in Late 19th Century Ontario."
90 *RAMCMP,* 1924–25, 370 (our translation).
91 E.T.D. Chambers, *The Ouananiche and Its Canadian Environment* (New York: Harper, 1896).
92 *RAMCMP,* 1930–31, 524.
93 Taylor's quotation in *RAMCMP,* 1932–33, 184 (our translation). On the change of vocation of the Magog fish nursery, see *RAMCMP,* 1933–34, 234.
94 V.C. Wynne-Edwards, "The Breeding Habits of the Black-Headed Minnow (*Pimephales promelas* Raf.)," *Transactions of the American Fisheries Society* 62 (1932): 382–83.
95 Vadim D. Vladykov, "Conditions de pêche à la truite dans le camp Jacques-Cartier, durant 1936–40," in *Rapport de la Station biologique de Montréal et de la Station biologique du parc des Laurentides pour l'année 1941, travaux du ministère de la Chasse et de la Pêche de Québec,* Institut de biologie, Université de Montréal, Fascicule 4: 421–42; *RAMMP,* 1937–38, 212–14.
96 Archives de l'Université de Montréal, Fonds Vianney Legendre, P0271/ séries A4/ dossier 8, letter from Gustave Prévost to deputy minister Louis-Arthur Richard, December 28, 1936. See also Gustave Prévost and Lucien Piché, "Observations on the Respiration of Trout Fingerlings and a New Method of Transporting Speckled Trout (*Salvelinus fontinalis*)," *Transactions of the American Fisheries Society* 68 (1939): 344–53.

Chapter 5: Regionalization and Specialization of Agricultural Production

1 Courville, *Quebec,* 144–51.
2 J.-C. Taché, *Rapport du comité spécial sur l'état de l'agriculture du Bas-Canada* (Toronto: Louis Perrault, 1850).
3 N. Perron, "Genèses des activités laitières," 113–40.
4 Hodgetts, *Pioneer Public Service,* 226–39; Fowke, *Canadian Agricultural Policy,* 112–16, 148–51.
5 N. Perron, *L'État et le changement agricole dans Charlevoix,* 91.
6 Hamelin and Roby, *Histoire économique du Québec,* 188.
7 Ibid., 189.
8 M.-A. Perron, *Un grand éducateur agricole,* 142–43.
9 N. Perron, *L'État et le changement agricole dans Charlevoix,* 109.
10 M.-A. Perron, *Un grand éducateur agricole,* 147–48; *Statuts de la province de Québec,* 56 Vict. 1893, c. 20, art. 1587.
11 *Journal d'agriculture,* July 15, 1918, 6, cited in N. Perron, *L'État et le changement agricole dans Charlevoix,* 124n17.

12 "Revue de l'année 1884," *Journal d'agriculture illustré* 8, 1 (January 1885): 2.
13 Gow, *Histoire de l'administration publique québécoise*, 51.
14 N. Perron, "Genèses des activités laitières," 119.
15 Hamelin and Roby, *Histoire économique du Québec*, 200.
16 N. Perron, *L'État et le changement agricole dans Charlevoix*, 97–98.
17 Jean, "Idéologies et professionnalisation," 252.
18 *RAMA*, 1912–13, v.
19 *RAMA*, 1915–16, xiii.
20 McGill University Archives, Macdonald College fonds (RG-43), vol. 4, dossier 194, Principal Harrison to Martin Burrell, federal Minister of Agriculture, June 27, 1912; Principal Harrison to P.S.G. Mackenzie, provincial Minister of Finance, May 14, 1913; dossier 195, letter from Principal Harrison to "Macdonald College Demonstrators," October 22, 1913; Snell, *Macdonald College of McGill University*, 169; Létourneau, *Histoire de l'agriculture*, 253. Financial difficulties at the university forced the college to end the demonstrations in 1916. RG-43, vol. 4, dossier 195, letter from Principal Harrison to "Macdonald College Demonstrators," February 1, 1916.
21 "Un agronome pour la région de l'Abitibi," *Journal d'agriculture et d'horticulture illustré* 18, 5 (November 15, 1914): 91.
22 BANQ-Québec, Fonds ministère de l'Agriculture, E9, S100, SS1, SSS2, bte 26, dossier 1649/22, "Établissement du service d'agronomie, ses buts."
23 *RAMA*, 1922–23, vi.
24 Guay, *Genèse de la régionalisation agricole*, 11 (our translation).
25 On agriculture in the districts of the Island of Montreal and the region around it, see S. Castonguay, "Agriculture and Urban Society on the Island of Montreal." For the Richelieu Valley, see Beauregard, "Les étapes de la mise en valeur agricole de la vallée du Richelieu."
26 *RAMA*, 1923–24, 135.
27 Courville, *Quebec*, 237.
28 BANQ–Rouyn-Noranda, Fonds ministère de l'Agriculture, des Pêcheries et de l'Alimentation. Direction régionale de l'Abitibi-Témiscamingue, Alexandre Rioux, agronome, E9, S3, 1982-06-004/1; 08-y, E9-3/1-3, Agronome – Travail, visites et déplacements, 1931–1932, aux agronomes, February 24, 1931, Narcisse Savoie, service chief.
29 *RAMA*, 1920–21, 129 (our translation). The ridings were those of Assomption, Terrebonne-Laval, Argenteuil, Deux-Montagnes, Vaudreuil-Soulanges, Laprairie-Napierville, Beauharnois, Huntingdon, Châteauguay, Chambly, Rouville, and Saint-Jean-d'Iberville.
30 Perron died in 1930, and a sixth district was created in March 1931 that included the ridings north of Montreal, but new principles guided a redrawing of the agronomic map in 1932. BANQ–Rouyn-Noranda, E9, S3, 1982-06-004/1; 08-y, E9-3/1-3, Agronome – Travail, visites et déplacements, 1931–1932, "Au personnel extérieur du ministère de l'Agriculture," July 4, 1933, L.-Ph. Roy, président du Conseil des chefs.
31 René Castonguay, "Joseph-Léonide Perron," in *Dictionary of Canadian Biography*, vol. 15 (Quebec City/Toronto: Université Laval/University of Toronto, 2003), http://www.biographi.ca/en/bio/perron_joseph_leonide_15E.html.
32 *RAMA*, 1927–28, iii (our translation). Perron's reform is discussed in the foreword of the new edition of Kesteman et al., *Histoire du syndicalisme agricole au Québec*.
33 Létourneau, *Histoire de l'agriculture*, 258.
34 *RAMA*, 1928–29, 100–1 (our translation). The three service heads were Louis-Philippe Roy, Stanislas-J. Chagnon, and a professor from the Institut d'Oka, Henri-Charles Bois, who later became the head of the rural economy service. The articles in the newsletter were Jean-Charles Magnan, "Dans le monde agricole: Arthur Fortin," *Le Lien* 1, 7 (September 1928): 99–101; Paul Gingras, "Programme d'action agricole," *Le Lien* 1, 7 (September 1928): 109–13. See also

Firmin Létourneau, "Ce que M. Perron pourrait faire s'il devenait ministre de l'Agriculture," *Le Bulletin des agriculteurs* 14, 16 (April 25, 1929).
35 *RAMA*, 1928–29, v (our translation).
36 Ibid., vii (our translation).
37 Ibid., viii–ix (our translation).
38 Ibid., v.
39 Jean-Charles Magnan, "Tout le monde à la tâche: Un nouveau chef. – Un programme. – Espoir et confiance!" *Le Lien* 2, 5 (May 1929): 25 (our translation).
40 Genest, *Godbout*, 59–60.
41 BANQ–Rouyn-Noranda, E9, S3, 1982-06-004/1; 08-y, E9-3/1-3, Agronome – Travail, visites et déplacements, 1931–1932, "Au personnel extérieur du ministère de l'Agriculture," July 4, 1933, L.-P. Roy, président du Conseil des chefs de service; *RAMA*, 1932–33, 3.
42 Frenette, "La recherche d'un cadre régional au Québec méridional."
43 For instance, districts 7, 9, and 10 exchanged the ridings of Bagot, Richelieu, and Rouville, depending on whether they figured in the description of the annual report or in the map that accompanied it. *Annuaire statistique de la province du Québec*, 22nd year, 1935, 290–95.
44 *Annuaire statistique de la province de Québec*, 1931, 231.
45 *Annuaire statistique de la province de Québec*, 1931–38, "Production" section.
46 On the normative dimension of agricultural statistics, see Murdoch and Ward, "Governmentality and Territoriality."
47 McGill University Archives, RG-43, vol. 14, dossier 598, letter from interim dean J.F. Snell to deputy minister of agriculture, January 13, 1913.
48 BANQ–Rouyn-Noranda, E9, S3, 1982-06-004/1; 08-y, E9-3/1-3, Agronome – Travail, visites et déplacements, 1931–1932, "Au personnel extérieur du ministère de l'Agriculture," July 4, 1933, L.-P. Roy, président du Conseil des chefs de service.
49 BANQ–Rouyn-Noranda, E9, S3, 1982-06-004/1; 08-y, E9-3/1-3, Agronome – Travail, visites et déplacements, 1931–1932, "Au personnel extérieur du ministère de l'Agriculture," May 3, 1933, Narcisse Savoie, chef du service administratif.
50 *RAMA*, 1932–33, 26.
51 Ibid., 3–4.
52 Guay, *Genèse de la régionalisation agricole*, 8 (our translation).
53 Albert Rioux, *Je me souviens: mémoires d'Albert Rioux* (Longueuil: Éditions de la Terre de chez nous, 1982), 63–65. On Caron, see Jacques Saint-Pierre, "Caron, Joseph-Édouard," in *Dictionary of Canadian Biography*, vol. 15 (Quebec City/Toronto: Université Laval/University of Toronto, 2003), http://www.biographi.ca/en/bio/caron_joseph_edouard_15E.html.
54 *En pleine décadence: l'agriculture sous M. Caron* (Montreal: Imprimerie populaire, 1926), 7. On the Department of Agriculture under Caron, see Louise Blondin, "La politique agricole au Québec," 102–24.
55 BANQ–Rouyn-Noranda, E9, S3, 1982-06-004/1; 08-y, E9-3/1-3, Agronome – Travail, visites et déplacements, 1931–1932, "Aux agronomes," August 3, 1931, Narcisse Savoie, chef du service.
56 N. Perron, "Genèses des activités laitières," 113–40; Thibeault, *Industrie laitière et transformation agraire au Saguenay–Lac-Saint-Jean*.
57 Gow, *Histoire de l'administration publique québécoise*, 50.
58 *RAMA*, 1928–29, iii–iv (our translation).
59 Martin, *Les fruits du Québec*.
60 Ibid., 87.
61 *Annual Report of the Pomological and Fruit Growing Society of the Province of Quebec* (Montreal: John Lovell, 1894), 111 (our translation).

62 BANQ-Québec, E9, S100, SS1, SSS2, box 8, dossier 575/97, Auguste Dupuis to premier of the province of Quebec, J.E. Flynn, March 1, 1897.
63 On the Société pomologique, see BANQ-Québec, E9, S100, SS1, SSS2, box 8, dossier 608/98, W.W. Dunlop, secretary of the Pomological and Fruit Growing Society of the Province of Quebec, "Some of the advantages to be derived from the establishment of fruit experimental stations in this province," [n.d.]. On the plant distribution site, see BANQ-Québec, E9, S100, SS1, SSS2, box 13, dossier 2913/02, "Rapport concernant les stations d'arboriculture fruitière," October 18, 1902.
64 On "reasoned exploitation," see BANQ-Québec, E9, S100, SS1, SSS2, box 8, dossier 575/97, Formule de contrat entre le département de l'Agriculture, Québec, et les cultivateurs qui se chargent de l'établissement et du maintien de petites stations expérimentales d'arboriculture fruitière [n.d.]. On the quotation "information useful ...," see Jean-Baptiste Roy, *Histoire de la pomologie au Québec* (Quebec City: Ministère de l'Agriculture, des Pêcheries et de l'Alimentation du Québec, 1978), 31 (our translation).
65 Roy, *Histoire de la pomologie* (our translation).
66 Ibid. (our translation).
67 "Culture fruitière: stations fruitières expérimentales: vergers de démonstration," *Journal d'agriculture et d'horticulture illustré* 15, 8 (February 15, 1912): 151–52 (our translation).
68 *RAMA*, 1913, 132.
69 Provencher, *La Station de recherche de Deschambault*.
70 S. Castonguay, *Protection des cultures, construction de la nature*, 54.
71 Létourneau, *Histoire de l'agriculture*, 254 (our translation).
72 Martin, *Les fruits du Québec*, 94 (our translation).
73 Roy, *Histoire de la pomologie*, 32.
74 Ibid.
75 *RAMA*, 1925, 185.
76 Roy, *Histoire de la pomologie*, 35.
77 See also Armand Létourneau, "Une chicane à l'exposition de pommes," *Journal d'agriculture* 30, 5 (November 25, 1926): 60.
78 *RAMA*, 1927–28, 140–41 (our translation).
79 Ibid., 140 (our translation).
80 Ibid., 138 (our translation).
81 Ibid., 141 (our translation).
82 "Le programme de l'honorable J.L. Perron, ministre de l'Agriculture: dans le domaine de l'horticulture," *Journal d'agriculture*, 33, 1 (July 25, 1929): 6 (our translation).
83 *RAMA*, 1927–28, 66 (our translation).
84 The Lower St. Lawrence region was added to these districts. "Pomiculteurs, dites-vous que le Service d'arrosage est à votre disposition," *Journal d'agriculture*, 38, 45 (May 11, 1935): 5.
85 *RAMA*, 1927–28, 164.
86 Létourneau, "Une chicane à l'exposition de pommes," 60.
87 *RAMA*, 1919–20, xx; Blake A. Campbell and Albert Gosselin, *Le commerce des fruits et des légumes frais dans la ville de Montréal* (Quebec City: Ministère de l'Agriculture, 1940).
88 Boyd, "Making Meat"; Derry, *Art and Science in Breeding*.
89 Union expérimentale des agriculteurs de Québec, *Premier rapport général* (La Trappe: Institut agricole d'Oka, 1911), 7–8.
90 Jean-Charles Magnan, "Frère Liguori," in *Le monde agricole* (Montreal: Presses libres, 1972), 46–47.
91 Archives de l'Université de Montréal, Fonds de l'Institut Agricole d'Oka, dossier du Frère Liguori E0082/G2, 0074, "L'Union Expérimentale des Agriculteurs de Québec: heureuse métamorphose de la vie agricole dans notre province ...," [n.d.].

92 Joseph-Édouard Caron, "Notre agriculture provinciale," *Journal d'agriculture et d'horticulture illustré* 17, 6 (December 15, 1913): 105–7.
93 *RAMA*, 1909–10, 43–46.
94 Victor Fortier, "L'industrie avicole dans la province de Québec," *Journal d'agriculture et d'horticulture illustré* 15, 8 (February 15, 1912): 166–67.
95 *RAMA*, 1909–10, 48–57 (our translation).
96 Archives de l'Université de Montréal, E0082/G2, 0074, "Rapport du RF Liguori et résumé du travail fait depuis septembre 1910 au 31 décembre 1911." See also Institut agricole d'Oka, "Le réveil avicole," *Journal d'agriculture et d'horticulture* 15, 6 (December 15, 1911): 118; *RAMA*, 1913–14, 139–40.
97 *RAMA*, 1913–14, vii.
98 See the report "Union expérimentale des agriculteurs de Québec," in *RAMA*, 1912–13, 110–11.
99 BANQ-Québec, E9 S100 SS1 SSS2, acc. 1960-01-029/21, Brother Liguori to J. Antonio Grenier, Secretary, Department of Agriculture, November 15, 1912; *RAMA*, 1913, 110–11.
100 *RAMA*, 1913–14, 108 (our translation).
101 *RAMA*, 1918–19, 159.
102 Jean-Baptiste Roy, *Histoire de l'aviculture au Québec* (Quebec City: Ministère de l'Agriculture, des Pêcheries et de l'Alimentation du Québec, 1978), 45 (our translation).
103 Union expérimentale des agriculteurs de Québec, "Élevage de la volaille," *Journal d'agriculture et d'horticulture* 14, 11 (May 15, 1911): 230 (our translation). See also *RAMA*, 1921–22, xi–xii.
104 Raoul Dumaine, "Choix des races de volailles," *Journal d'agriculture et d'horticulture illustré* 19, 2 (August 15, 1915): 34; *RAMA*, 1915–16, 13; Derry, *Art and Science in Breeding*.
105 Roy, *Histoire de l'aviculture*, 45 (our translation).
106 BANQ-Québec, E9, S100, SS1, SSS2, box 35, 3607, 30, J.-Antonio Grenier, Quebec deputy minister of agriculture, to John H. Grisdale, federal deputy minister of agriculture, March 6, 1930.
107 J.L. Perron, "Le programme de l'honorable J.L. Perron, ministre de l'Agriculture: l'aviculture," *Journal d'agriculture* 33, 1 (July 25, 1929): 5 (our translation).
108 *RAMA*, 1928–29, 54–56 (our translation).
109 Archives de l'Université de Montréal, E0082/G2, 0074, "L'Union expérimentale des agriculteurs de Québec: heureuse métamorphose de la vie agricole dans notre province..." [n.d.]. See also "Nouvelles avicoles: la production totale des couvoirs en 1935," *Journal d'agriculture* 39, 17 (October 26, 1935): 35.
110 For example, it was a member of the federal experimental farm service, Victor Fortier, who wrote the poultry columns in the *Journal d'agriculture illustré* until the late 1910s.

APPENDIX: IDENTIFICATION OF TECHNOSCIENTIFIC ACTIVITIES
IN THE PUBLIC ACCOUNTS

1 Dupré, "Un siècle de finances publiques québécoises," 562 (our translation).

Bibliography

PRIMARY SOURCES

Archives
Archives conserved at the Ministère des Ressources naturelles du Québec, Pépinière Proulx
Archives conserved at the Ministère des Ressources naturelles du Québec, Pépinière provinciale de Berthierville
Archives de l'École polytechnique de Montréal (AÉPM)
Archives de l'Université de Montréal
Archives du Séminaire de Québec (ASQ)
Bibliothèque et Archives nationales du Québec (BANQ), Montréal (BANQ-Montréal), Fonds de la Consolidated Bathurst
BANQ-Québec, Fonds ministère de l'Agriculture du Québec
BANQ-Québec, Fonds ministère de l'Énergie et des Ressources
BANQ-Québec, Fonds ministère des Terres et Forêts
BANQ-Québec, Fonds ministère du Loisir, de la Chasse et de la Pêche
BANQ–Rouyn-Noranda, Fonds ministère de l'Agriculture, des Pêcheries et de l'Alimentation
BANQ-Saguenay, Fonds Archives nationales du Québec, Cartes
McGill University Archives

Government Publications
Annuaire statistique de la province de Québec
État des comptes publics de la province de Québec
Geological Survey of Canada, Annual Reports
Geological Survey of Canada, Operations Reports
Journal d'agriculture
Journal d'agriculture et d'horticulture illustré
Journal d'agriculture illustré
Opérations minières dans la province de Québec
Rapport annuel du commissaire des Terres de la Couronne (*RACTC*)
Rapport annuel du commissaire des Terres de la Couronne, des Mines et des Pêcheries (*RACTCMP*)

Rapport annuel du commissaire des Terres, des Forêts et des Pêcheries (RACTFP)
Rapport annuel du ministère de l'Agriculture (RAMA)
Rapport annuel du ministère de la Colonisation et des Mines (RAMCM)
Rapport annuel du ministère de la Colonisation et des Pêcheries (RAMCP)
Rapport annuel du ministère de la Colonisation, des Mines et des Pêcheries (RAMCMP)
Rapport annuel du ministère de la Marine et des Pêcheries du Canada (RAMMPC)
Rapport annuel du ministère des Mines (RAMM)
Rapport annuel du ministère des Mines et des Pêcheries (RAMP)
Rapport annuel du ministère des Terres et Forêts (RAMTF)
Rapport annuel du ministère des Terres, des Forêts et des Pêcheries (RAMTFP)
Rapport annuel du ministère des Terres, des Mines et des Pêcheries (RAMTMP)
Rapport annuel du ministère des Terres, des Pêcheries, de la Colonisation (RAMTPC)

Periodicals
American Fisheries Society, Transactions
Canadian Forestry Association, Annual Report
Canadian Forestry Association, Report of the Annual Meeting
Canadian Forestry Journal
Canadian Mining and Mechanical Review
Canadian Mining Institute, Journal
Canadian Mining Institute, Transactions
Canadian Mining Journal
Forestry Quarterly
Illustrated Canadian Forest and Outdoors
La forêt et la ferme
La vie forestière
La vie forestière et rurale
Le Bulletin des agriculteurs
Le Lien
Pomological and Fruit Growing Society of the Province of Quebec, Annual Report
Revue trimestrielle canadienne
Rod and Gun Canada
Société de géographie du Québec, Bulletin

SECONDARY SOURCES

Alliès, Paul. *L'invention du territoire*. Grenoble: Presses universitaires de Grenoble, 1980.
Anderson, Benedict. *Imagined Communities: Reflections on the Origin and Spread of Nationalism*. Revised ed. New York: Verso, 1991. First published in 1983.
Anderson, Peter G. "Field Experiments: Critical Historical Geographies of Canada's Central Experimental Farm, 1886–1938." PhD dissertation, Queen's University, 2017.
Anholt, Dennis Munroe. "An Administrative History of the British Columbia Government Agents." PhD dissertation, University of Victoria, 1991.
Armstrong, Christopher, and H.V. Nelles. "Contrasting Development of the Hydro-Electric Industry in the Montreal and Toronto Regions, 1900–1930." *Journal of Canadian Studies* 18, 1 (1983): 5–27.
Armstrong, Robert. *Structure and Change: An Economic History of Quebec*. Toronto: Gage, 1984.
Audard, Catherine. *Qu'est-ce que le libéralisme? éthique, politique, société*. Paris: Gallimard, 2009.

Auger, Jean-François. "La recherche utilitaire dans les facultés de génie canadiennes: au service de l'industrie et du gouvernement, 1870–1950." PhD dissertation, Université du Québec à Montréal, 2004.

Bachand, Gilles. *Histoire de l'École de laiterie de Saint-Hyacinthe 1892–1985*. Belœil: Société d'histoire et de généalogie des Quatre Lieux, 2012.

Balogh, Brian. "Scientific Forestry and the Roots of the Modern American State: Gifford Pinchot's Path to Progressive Reform." *Environmental History* 7, 2 (2002): 198–225.

Beaud, Jean-Pierre, and Jean-Guy Prévost. "Models for Recording Age in Pre-1867 Canada: The Political-Cognitive Functions of Census Statistics." *Scientia Canadensis* 18 (1994): 135–50.

Beauregard, Ludger. "Les étapes de la mise en valeur agricole de la vallée du Richelieu." *Cahiers de géographie du Québec* 14, 32 (1970): 171–215.

Bellavance, Claude. "L'État, la 'houille blanche' et le grand capital: l'aliénation des ressources hydrauliques du domaine public québécois au début du XXe siècle." *Revue d'histoire de l'Amérique française* 51, 1 (1998): 1–32.

–. *Shawinigan Water and Power, 1898–1963: formation et déclin d'un groupe industriel au Québec*. Montreal: Boréal, 1994.

Biays, Pierre. "L'œkoumène agricole au Lac Saint-Jean." *Cahiers de géographie du Québec* 7, 13 (1962): 101–10.

Biggs, M. "Putting the State on the Map: Cartography, Territory, and European State Formation." *Comparative Studies in Society and History* 41, 2 (1999): 374–405.

Blanchard, Raoul. *Le Canada français: Province de Québec*. Montreal: Librarie Arthème Fayard, 1960.

Blondin, Louise. "La politique agricole au Québec, 1852–1929: le rôle du ministère de l'Agriculture dans l'évolution de cette industrie." Master's thesis, Université de Sherbrooke, 1987.

Boisclair, Guy. "Étude d'un mouvement de modernisation de l'agriculture: les premières années de l'Union catholique des cultivateurs dans le diocèse de Joliette, 1924–1952." PhD dissertation, Université du Québec à Trois-Rivières, 2002.

Bouchard, Gérard. *Quelques Arpents d'Amérique: population, économie, famille au Saguenay 1838–1971*. Montreal: Boréal, 1996.

Boudreau, Claude. *La cartographie au Québec: 1760–1840*. Sainte-Foy: Presses de l'Université Laval, 1994.

Boudreau, Claude, Serge Courville, and Normand Séguin. *Le territoire*. Sainte-Foy: Presses de l'Université Laval, 1997.

Bouthillier, Luc. "Le concept de rendement soutenu en foresterie dans un contexte nord-américain." PhD dissertation, Université Laval, 1991.

Boyd, William. "Making Meat: Science, Technology, and American Poultry Production." *Technology and Culture* 42, 4 (2001): 631–64.

Braun, Bruce. "Producing Vertical Territory: Geology and Governmentality in Late Victorian Canada." *Ecumene* 7, 1 (2000): 7–46.

Brenner, Neil, Bob Jessop, Martin Jones, and Gordon Macleod, eds. *State/Space: A Reader*. Oxford: Blackwell, 2003.

Brownstein, David. "Spasmodic Research as Executive Duties Permit: Space, Practice, and the Localization of Forest Management Expertise in British Columbia, 1912–1928." *Journal of Historical Geography* 52 (2016): 36–47.

Brun, Henri. "Le territoire du Québec: à la jonction de l'histoire et du droit constitutionnel." *Les Cahiers de droit* 33, 3 (1992): 927–43.

Brunelle, Richard. *Histoire de l'administration québécoise: le secteur minier*. Montreal: Université de Montréal, Département de sciences politiques, 1972.

Carroll, Patrick. *Science, Culture, and Modern State Formation*. Berkeley: University of California Press, 2006.

Carter, Paul. *The Road to Botany Bay: An Exploration of Landscape and History*. Minneapolis: University of Minnesota Press, 2010. First published 1987 by Faber and Faber.

Castonguay, René. *Rodolphe Lemieux et le Parti libéral, 1866–1937: Le chevalier du roi*. Sainte-Foy: Presses de l'Université Laval, 2000.

Castonguay, Stéphane. "Agriculture and Urban Society on the Island of Montreal." In *Montreal: The History of a North American City*, edited by Dany Fougères and Roderick MacLeod, vol. 1, 561–603. Montreal and Kingston: McGill-Queen's University Press, 2018.

–. "Connaissance et contrôle du milieu: les ressources de l'État moderne." In *Temps, espace et modernités*, edited by Brigitte Caulier and Yvan Rousseau, 295–307. Sainte-Foy: Presses de l'Université Laval, 2009.

–. "Faire du Québec un objet de l'histoire environnementale." *Globe* 9, 1 (2006): 17–49.

–. "Fédéralisme et centralisation de la recherche agricole au Canada: dynamique scientifique et écologie institutionnelle." *Bulletin d'histoire politique* 7, 3 (1999): 21–39.

–. "Foresterie scientifique et reforestation: l'État et la production d'une 'forêt à pâte' au Québec." *Revue d'histoire de l'Amérique française* 60, 1–2 (2006): 61–93.

–. "Naturalizing Federalism: Insect Outbreaks and the Centralization of Entomological Research in Canada, 1884–1914." *Canadian Historical Review* 85, 1 (2004): 1–34.

–. *Protection des cultures, construction de la nature: agriculture, foresterie et entomologie au Canada 1884–1959*. Sillery: Septentrion, 2004.

Castonguay, Stéphane, and Darin Kinsey. "The Nature of the Liberal Order Framework: State Formation, Conservation and the Government of Non-Humans in Canada." In *Liberalism and Hegemony: Debating the Canadian Liberal Revolution*, edited by Michel Ducharme and Jean-François Constant, 221–45. Toronto: University of Toronto Press, 2009.

Charland, Jean-Pierre. *Les pâtes et papiers au Québec: 1880–1980: Technologies, travail et travailleurs*. Quebec City: Institut québécois de recherche sur la culture, 1990.

Chartrand, Luc, Raymond Duchesne, and Yves Gingras. *Histoire des sciences au Québec: de la Nouvelle-France à nos jours*. 2nd ed. Montreal: Boréal, 2009. First published 1987 by Boréal.

Cooper, Alix. "'The Possibilities of the Land': The Inventory of 'Natural Riches' in the Early Modern German Territories." *History of Political Economy* 35, 5 (2003): 129–53.

Côté, Gaston. "Domestiquer le sauvage: chasseurs sportifs et gestion de la grande faune au Québec (1858–2004)." PhD dissertation, Université du Québec à Trois-Rivières, 2017.

Courville, Serge. *Immigration, colonisation et propagande: du rêve américain au rêve colonial*. Sainte-Foy: Multimondes, 2002.

Courville, Serge. *Quebec: A Historical Geography*. Translated by Richard Howard. Vancouver: UBC Press, 2008.

Cox, Kevin R. "Political Geography and the Territorial." *Political Geography* 22, 6 (2003): 607–10.

–. *Political Geography: Territory, State, and Society*. Oxford: Blackwell, 2002.

Craib, Raymond B. *Cartographic Mexico: A History of State Fixations and Fugitive Landscapes*. Durham, NC: Duke University Press, 2004.

Curtis, Bruce. *The Politics of Population: State Formation, Statistics, and the Census of Canada, 1840–1875*. Toronto: University of Toronto Press, 2000.

De Vecchi, Vittorio Maria Giuseppe. "Science and Government in Nineteenth-Century Canada." PhD dissertation, University of Toronto, 1978.

Demeritt, David. "Scientific Forest Conservation and the Statistical Picturing of Nature's Limits in the Progressive-Era United States." *Environment and Planning D: Society and Space* 19, 4 (2001): 431–59.

Derry, Margaret. *Art and Science in Breeding: Creating Better Chickens*. Toronto: University of Toronto Press, 2012.
Deschênes, Gaston. *Dictionnaire des parlementaires du Québec, 1792–1992*. Sainte-Foy: Presses de l'Université Laval, 1993.
Doern, Bruce. *The National Research Council in the Innovation Policy Era: Changing Hierarchies, Networks, and Markets*. Toronto: University of Toronto Press, 2002.
Dorion, Henri, and Jean-Paul Lacasse. *Le Québec: territoire incertain*. Sillery: Septentrion, 2011.
Douglas, Mary. *How Institutions Think*. Syracuse, NY: Syracuse University Press, 1986.
Driver, Felix. "Political Geography and State Formation: Disputed Territory." *Progress in Human Geography* 15, 3 (1991): 268–80.
Drolet, Jean-Paul. "La Loi des mines de Québec." *Revue du barreau* 9, 2 (1949): 136–49.
Dunlap, Thomas R. *Saving America's Wildlife: Ecology and the American Mind, 1850–1950*. Princeton, NJ: Princeton University Press, 1988.
Dupré, Ruth. "The Evolution of Quebec Government Expenditures, 1867–1969." PhD dissertation, University of Toronto, 1987.
–. "Regulating the Quebec Dairy Industry, 1905–1921: Peeling Off the Joseph Label." *Journal of Economic History* 50, 2 (1990): 339–48.
–. "Un siècle de finances publiques québécoises: 1867–1969." *L'Actualité économique* 64, 4 (1988): 559–83.
Dupree, A. Hunter. *Science in the Federal Government: A History of Policies and Activities*. 2nd ed. Baltimore: Johns Hopkins University Press, 1986. First published 1957 by Belknap Press of Harvard University Press.
Dutil, Patrice. *L'avocat du diable: Godfroy Langlois et la politique du libéralisme progressiste à l'époque de Laurier*. Montreal: Éditions Robert Davies, 1995.
Edney, Matthew H. *Mapping an Empire: The Geographical Construction of British India, 1765–1843*. Chicago: University of Chicago Press, 1997.
Elden, Stuart. "Governmentality, Calculation, Territory." *Environment and Planning D: Society and Space* 25, 3 (2007): 562–80.
–. "Rethinking Governmentality." *Political Geography* 26, 1 (2007): 29–33.
Elden, Stuart, and Jeremy W. Crampton, eds. *Space, Knowledge and Power: Foucault and Geography*. Aldershot, UK: Ashgate, 2007.
Escolar, Marcelo. "Exploration, cartographie et modernisation du pouvoir de l'État." *Revue internationale des sciences sociales* 151 (1997): 59–78.
Evans, Peter B., Dietrich Rueschemeyer, and Theda Skocpol, eds. *Bringing the State Back In*. Cambridge: Cambridge University Press, 1985.
Flamand-Hubert, Maude. "La forêt québécoise en discours dans la première moitié du XXe siècle: représentations politiques et littéraires." PhD dissertation, Université du Québec à Rimouski and Université Paris-Sorbonne, 2017.
Foisy-Geoffroy, Dominique. *Esdras Minville: nationalisme économique et catholicisme social au Québec durant l'entre-deux-guerres*. Sillery: Septentrion, 2004.
Fortier, Robert, ed. *Villes industrielles planifiées*. Montreal: Boréal, 1996.
Foucault, Michel. *Discipline and Punish: The Birth of the Prison*. 1975. Reprint, New York: Vintage, 1995.
–. *Sécurité, territoire, population*. Paris: Gallimard, 2004.
Fowke, Vernon C. *Canadian Agricultural Policy: The Historical Pattern*. Toronto: University of Toronto Press, 1946.
Frenette, Jean-Vianney. "La recherche d'un cadre régional au Québec méridional: quelques étapes, de 1932 à 1966." *Cahiers de géographie du Québec* 17, 40 (1973): 69–84.
Gagnon, Serge. "L'appropriation ludique de la forêt au Québec: d'une gestion privée de 'club' à une intervention publique de protection (1885–1935)." Special issue, "Espaces et aires

protégés: gestion intégrée et gouvernance participative," edited by Pascal Saffache. *Études caribéennes* 12 (2009): 119–40.

–. *L'échiquier touristique québécois*. Sainte-Foy: Presses de l'Université du Québec, 2003.

Gallichan, Gilles. *Honoré Mercier: la politique et la culture*. Sillery: Septentrion, 1994.

Gaudreau, Guy. "L'État, le mesurage du bois et la promotion de l'industrie papetière." *Revue d'histoire de l'Amérique française* 43, 2 (1989): 203–19.

–. "Exploitation des forêts publiques au Québec (1874–1905): transition et nouvel essor." *Revue d'histoire de l'Amérique française* 42, 1 (1988): 3–27.

–. *L'histoire des mineurs du Nord ontarien et québécois*. Sillery: Septentrion, 2003.

–. *Les récoltes des forêts publiques au Québec et en Ontario: 1840–1900*. Montreal and Kingston: McGill-Queen's University Press, 1999.

Gélinas, Cyrille. *L'enseignement et la recherche en foresterie à l'Université Laval de 1910 à nos jours*. Quebec City: Société d'histoire forestière du Québec, 2010.

Genest, Jean-Guy. *Godbout*. Sillery: Septentrion, 1996.

Gibbons, Michael, Camille Limoges, Helga Nowotny, Simon Schwartzman, Peter Scott, and Martin Trow. *The New Production of Knowledge: The Dynamics of Science and Research in Contemporary Societies*. London: Sage, 1994.

Gillis, R. Peter, and T.R. Roach. *Lost Initiatives: Canada's Forest Industries, Forest Policy and Forest Conservation*. New York: Greenwood Press, 1986.

Gingras, Sylvain. *A Century of Sport: Hunting and Fishing in Quebec*. Saint-Raymond: Éditions Rapides blancs, 1994.

Girard, Camil, and Normand Perron. *Histoire du Saguenay–Lac-Saint-Jean*. Quebec City: IQRC, 1989.

Girard, Michel F. "La forêt dénaturée: les discours sur la conservation de la forêt québécoise au tournant du XXe siècle." Master's thesis, University of Ottawa, 1988.

Gottman, Jean. *La politique des États et leur géographie*. Paris: Éditions du CTHS, 2007. First published 1952 by Librairie Armand Colin.

Gow, James Iain. "L'histoire de l'administration publique québécoise." *Recherches sociographiques* 16, 3 (1975): 385–411.

–. *Histoire de l'administration publique québécoise, 1867–1970*. Montreal: Presses de l'Université de Montréal, 1986.

Grek-Martin, Jason William. "Making Settler Space: George Dawson, the Geological Survey of Canada and the Colonization of the Canadian West in the Late 19th Century." PhD dissertation, Queen's University, 2009.

–. "Survey Science on Trial: The Geographic Contours of Geology's Practical Science Debate in Late Victorian Canada." *Journal of Historical Geography* 45 (2014): 1–11.

Guay, Donald. *Genèse de la régionalisation agricole (1913–1993)*. [Quebec City]: Ministère de l'Agriculture, des Pêcheries et de l'Alimentation du Québec, [1993].

–. *Histoires vraies de la chasse*. Montreal: VLB, 1983.

Guha, Ramachandra. *The Unquiet Woods: Ecological Change and Peasant Resistance in the Himalaya*. Oxford: Oxford University Press, 1992.

Häkli, Jouni. "Territoriality and the Rise of the Modern State." *Fennia* 172, 1 (1994): 1–82.

Hamel, Thérèse, Michel Morisset, and Jacques Tondreau. *De la terre à l'école: histoire de l'enseignement agricole au Québec, 1926–1969*. Montreal: Hurtubise HMH, 2000.

Hamelin, Jean. *Histoire de l'Université Laval: Les péripéties d'une idée*. Sainte-Foy: Presses de l'Université Laval, 1995.

Hamelin, Jean, and Yves Roby. *Histoire économique du Québec, 1851–1896*. Montreal: Fides, 1971.

Hannah, Matthew G. *Governmentality and the Mastery of Territory in Nineteenth-Century America*. Cambridge: Cambridge University Press, 2000.

Hardy, René. "Exploitation forestière et environnement au Québec, 1850–1920." *Zeitschrift für Kanada-Studien* 15, 27 (1995): 63–79.
Hardy, René, and Normand Séguin. *Forêt et société en Mauricie: la formation de la région de Trois-Rivières, 1830–1930*. Montreal: Boréal Express, 1984.
–. *Histoire de la Mauricie*. Sainte-Foy: IQRC, 2004.
Harley, J.B. "Maps, Knowledge and Power." In *The Iconography of Landscape: Essays on the Symbolic Representation, Design and Use of Past Environments*, edited by Denis Cosgrove and Stephen Daniels, 277–312. Cambridge: Cambridge University Press, 1988.
–. *The New Nature of Maps: Essays in the History of Cartography*. Baltimore: Johns Hopkins University Press, 2001.
Harris, Robin S. *A History of Higher Education in Canada, 1663–1960*. Toronto: University of Toronto Press, 1976.
Hays, Samuel P. *Conservation and the Gospel of Efficiency: The Progressive Conservation Movement, 1890–1920*. Cambridge, MA: Harvard University Press, 1959.
Heaman, Elsbeth. *The Inglorious Arts of Peace: Exhibitions in Canadian Society during the Nineteenth Century*. Toronto: University of Toronto Press, 1999.
Hébert, Yves. "Conservation, culture et identité: la création du parc des Laurentides et du parc de la Montagne-Tremblante, 1894–1938." In *Changing Parks: The History, Future, and Cultural Context of Parks and Heritage Landscapes*, edited by John S. Marsh and Bruce W. Hodgins, 140–59. Toronto: Natural Heritage and Natural History, 1998.
–. *Une histoire de l'écologie au Québec: les regards sur la nature des origines à nos jours*. Quebec City: Éditions GID, 2006.
Hodgetts, J.E. *From Arm's Length to Hands On: The Formative Years of Ontario's Public Service*. Toronto: University of Toronto Press, 1995.
–. *Pioneer Public Service: An Administrative History of the United Canadas, 1841–1867*. Toronto: University of Toronto Press, 1955.
Hodgetts, J.E., William McCloskey, Reginald Whitaker, and V. Seymour Wilson. *The Biography of an Institution: The Civil Service Commission of Canada, 1908–1967*. Montreal and Kingston: McGill-Queen's University Press, 1972.
Hodgins, Bruce W., Jamie Benidickson, and Peter Gillis. "The Ontario and Quebec Experiment in Forest Reserves 1883–1930." *Journal of Forest History* 26, 1 (1982): 20–33.
Huxley, Margo. "Geographies of Governmentality." In *Space, Knowledge and Power: Foucault and Geography*, edited by Jeremy W. Crampton and Stuart Elden, 185–204. Aldershot, UK: Ashgate, 2007.
–. "Spatial Rationalities: Order, Environment, Evolution and Government." *Social and Cultural Geography* 7, 5 (2006): 771–87.
Immarigeon, Henriette. "Les frontières du Québec." In *Le territoire québécois*, edited by Jacques Brossard, 5–47. Montreal: Presses de l'Université de Montréal, 1970.
Ingold, Alice. "Naming and Mapping National Resources in Italy (19th Century): Propositions for a History of Categorizing 'Natural Resources.'" In *Views from the South. Environmental Stories from the Mediterranean World (19th–20th Centuries)*, edited by Marco Armiero, 51–65. Naples: Consiglio Nazionale delle Ricerche, 2006.
Ingram, Darcy. "'Au temps et dans les quantités qui lui plaisent': Poachers, Outlaws, and Rural Banditry in Quebec." *Histoire sociale* 42, 83 (2009): 1–34.
–. *Wildlife, Conservation, and Conflict in Quebec, 1840–1914*. Vancouver: UBC Press, 2013.
Isenberg, Andrew. *The Destruction of the Bison: An Environmental History, 1750–1920*. Cambridge: Cambridge University Press, 2001.
Jacoby, K. *Crimes against Nature: Squatters, Poachers, Thieves, and the Hidden History of American Conservation*. Berkeley: University of California Press, 2001.

Jarrell, Richard A. *The Cold Light of Dawn: A History of Canadian Astronomy.* Toronto: University of Toronto Press, 1988.
Jean, Bruno. "Idéologies et professionnalisation: le cas des agronomes." *Recherches sociographiques* 9, 2 (1978): 251–60.
Johnston, Ron J. "Out of the Moribund Backwater: Territory and Territoriality in Political Geography." *Political Geography* 20, 6 (2001): 677–93.
–. "Territoriality and the State." In *Geography, History and Social Sciences,* edited by Georges B. Benko and U. Strohmayer, 213–25. Dordrecht, Netherlands: Kluwer, 1995.
Johnstone, Kenneth. *Forêts et tourments: 75 ans d'histoire du Service fédéral des forêts 1899–1974.* Ottawa: Forêts Canada, 1991.
Joyce, Patrick. *The Rule of Freedom: Liberalism and the Modern City.* New York: Verso, 2003.
Kesteman, Jean-Pierre, Guy Boisclair, Jean-Marc Kirouac, and Jocelyn Morneau. *Histoire du syndicalisme agricole au Québec: UCC-UPA, 1924–2004.* 2nd ed. Montreal: Boréal, 2004. First published 1984.
Kinsey, Darin S. "Fashioning a Freshwater Eden: Elite Anglers, Fish Culture, and State Development of Québec's 'Sport' Fishery." PhD dissertation, Université du Québec à Trois-Rivières, 2008.
–. "'Seeding the Water as the Earth': The Epicenter and Peripheries of a Western Aquacultural Revolution." *Environmental History* 11, 3 (2006): 527–66.
Kirsch, Scott. "John Wesley Powell and the Mapping of the Colorado Plateau, 1869–1879: Survey Science, Geographical Solutions, and the Economy of Environmental Values." *Annals of the Association of American Geographers* 92, 5 (2002): 548–72.
Knight, William. "Samuel Wilmot, Fish Culture, and Recreational Fisheries in Late 19th Century Ontario." *Scientia Canadensis* 30, 1 (2007): 75–90.
Kuhlberg, Mark. *One Hundred Rings and Counting: Forestry Education and Forestry in Toronto and Canada, 1907–2007.* Toronto: University of Toronto Press, 2009.
Lacasse, Jean-Paul. "La propriété des mines en droit québécois." *Justinien* 2 (1965): 22–41.
–. "Réserve des trois chaînes et gestion du domaine public foncier au Québec." *Revue générale de droit* 8, 1 (1977): 101–10.
Lambert, R.S., and A.P. Pross. *Renewing Nature's Wealth: A Centennial History of the Public Management of Lands, Forests and Wildlife in Ontario.* Toronto: Department of Lands and Forests, 1967.
Latour, Bruno. *Science in Action: How to Follow Scientists and Engineers through Society.* Cambridge, MA: Harvard University Press, 1987.
–. *We Have Never Been Modern.* Cambridge, MA: Harvard University Press, 1993.
Lefebvre, Henri. *De l'État,* vol. 4, *Les contradictions de l'État moderne: la dialectique et/de l'État.* Paris: Maspero, 1978.
Legg, Stephen. "Foucault's Population Geographies: Classifications, Biopolitics and Governmental Spaces." *Population, Space and Place* 11, 3 (2005): 137–56.
Létourneau, Firmin. *Histoire de l'agriculture (Canada français).* 2nd ed. Montreal: Fides, 1952. First published 1950.
Leuenberger, Christine, and Izhak Schnell. "The Politics of Maps: Constructing National Territories in Israel." *Social Studies of Science* 40, 6 (2010): 803–42.
Lévesque, Ulric, Denis Dumont, Maurice Proulx, Fondation François-Pilote, et Institut de technologie agro-alimentaire [La Pocatière, QC]. *150 ans d'enseignement agricole à La Pocatière (1859–2009),* vol. 1, *L'école et la faculté (1859–1962).* La Pocatière, QC: Fondation François-Pilote, 2009.
Linteau, Paul-André, René Durocher, and Jean-Claude Robert. *Quebec: A History 1867–1929.* Translated by Robert Chodos. Toronto: James Lorimer, 1983.

Little, Jack. *Colonialism, Nationalism and Capitalism: The Upper St. Francis.* Montreal and Kingston: McGill-Queen's University Press, 1988.

Lowood, H. "The Calculating Forester: Quantification, Cameral Science, and the Emergence of Scientific Forestry Management in Germany." In *The Quantifying Spirit in the 18th Century,* edited by T. Frängsmyr, J. Heilbron, and R. Rider, 315–42. Berkeley: University of California Press, 1990.

Mann, Michael. "The Autonomous Power of the State: Its Origins, Mechanisms and Results." *Archives européennes de sociologie* 25, 2 (1984): 185–213.

Marr, William, and Donald Paterson. *Canada: An Economic History.* Toronto: Macmillan, 1980.

Martin, Paul-Louis. *La chasse au Québec.* Montreal: Boréal, 1990.

–. *Les fruits du Québec: histoire et traditions des douceurs de la table.* Sillery: Septentrion, 2002.

Massell, David. *Amassing Power: J.B. Duke and the Saguenay River, 1897–1927.* Montreal and Kingston: McGill-Queen's University Press, 2000.

–. *Quebec Hydropolitics: The Peribonka Concessions of the Second World War.* Montreal and Kingston: McGill-Queen's University Press, 2011.

Maxwell, J.A. *Federal Subsidies to the Provincial Governments in Canada.* Cambridge, MA: Harvard University Press, 1937.

McDonald, Robert A.J. "The Quest for 'Modern Administration': British Columbia's Civil Service, 1870s to 1940s." *BC Studies* 161 (2009): 9–35.

McGee, Jean-Charles. *Le département des Terres de la Couronne à partir de 1763.* [Quebec City]: Ministère de la Fonction publique, 1974.

McRoberts, Kenneth, and Dale Postgate. *Développement et modernisation du Québec.* Montreal: Boréal, 1983.

Minville, Esdras. *L'Agriculture: étude préparée avec la collaboration de l'Institut agricole d'Oka.* Montreal: Fides, 1943.

–. *La Forêt: étude préparée avec la collaboration de l'École de génie forestier de Québec.* Montreal: Fides, 1944.

–. *Notre milieu: aperçu général sur la province de Québec.* Montreal: Fides, 1942.

–. *Pêche et chasse: étude préparée avec la collaboration du Département des pêcheries maritimes et du Département de la chasse et de la pêche de Québec, ainsi que de l'École supérieure des pêcheries de Sainte-Anne-de-la-Pocatière.* Montreal: Fides, 1946.

Morissonneau, Christian. *La Société de géographie du Québec, 1877–1970.* Quebec City: Presses de l'Université Laval, 1971.

Mukerji, Chandra. *A Fragile Power: Scientists and the State.* Princeton, NJ: Princeton University Press, 1989.

–. "The Great Forestry Survey of 1669–1671: The Use of Archives for Political Reform." *Social Studies of Science* 37 (2007): 227–53.

–. "Intelligent Uses of Engineering and the Legitimacy of Power." *Technology and Culture* 44, 4 (2003): 655–76.

–. "The Political Mobilization of Nature in Seventeenth-Century French Formal Gardens." *Theory and Society* 23, 5 (1994): 651–77.

–. *Territorial Ambitions and the Gardens of Versailles.* Cambridge: Cambridge University Press, 1997.

Murdoch, Jonathan, and Neil Ward. "Governmentality and Territoriality: The Statistical Manufacture of Britain's National Farm." *Political Geography* 16, 4 (1997): 307–32.

Murphy, Alexander B. "Entente Territorial: Sack and Raffestin on Territoriality." *Environment and Planning D: Society and Space* 30, 1 (2012): 157–72.

Nelles, H.V. *The Politics of Development: Forests, Mines, and Hydro-Electric Power in Ontario, 1849–1941.* Toronto: Macmillan, 1974.

Olson, Sherry H. *The Depletion Myth: A History of Railroad Use of Timber.* Cambridge, MA: Harvard University Press, 1971.
Otis, Yves. "La différenciation des producteurs laitiers et le marché de Montréal (1900–1930)." *Revue d'histoire de l'Amérique française* 45, 1 (1991): 39–71.
Otter, Chris. *The Victorian Eye: A Political History of Light and Vision in Britain, 1800–1910.* Chicago: University of Chicago Press, 2008.
Owram, Doug. *Promise of Eden: The Canadian Expansionist Movement and the Idea of the West, 1856–1900.* Toronto: University of Toronto Press, 1980.
Paasi, Anssi. "Bounded Spaces in a 'Borderless World': Border Studies, Power and the Anatomy of Territory." *Journal of Power* 2, 2 (2009): 213–34.
Painter, Joe, and Alexander Sam Jeffrey. *Political Geography: An Introduction to Space and Power.* London: Sage, 2009.
Paquette, Pierre. "Industries et politiques minières au Québec: une analyse économique 1896–1975." *Revue d'histoire de l'Amérique française* 37, 4 (1984): 593–97.
–. *Les mines du Québec, 1867–1975: une évaluation critique d'un mode historique d'industrialisation nationale.* Outremont: Carte blanche, 2000.
Pauly, Philip. *Biologists and the Promise of American Life: From Meriwether Lewis to Alfred Kinsey.* Princeton, NJ: Princeton University Press, 2002.
Peluso, Nancy. *Rich Forest, Poor People: Resource Control and Resistance in Java.* Berkeley: University of California Press, 1992.
Pepin, Michel. *Histoire et petites histoires des vétérinaires du Québec.* Montreal: Éditions François Lubrina, 1986.
Père Louis-Marie, OCSO. *L'Institut d'Oka: cinquantenaire, 1893–1943: École agricole, Institut agronomique.* [La Trappe]: [Institut agricole d'Oka], [1944].
Perron, Marc-A. *Un grand éducateur agricole, Édouard-A. Barnard, 1835–1898: étude historique sur l'agriculture de 1760 à 1900.* Quebec City: self-published, 1955.
Perron, Normand. *L'État et le changement agricole dans Charlevoix, 1850–1950.* Sainte-Foy: Presses de l'Université Laval, 2003.
–. "Genèses des activités laitières, 1850–1960." In *Agriculture et colonisation au Québec: aspects historiques,* edited by Normand Séguin, 113–40. Montreal: Boréal Express, 1980.
Pestre, Dominique. *Science, argent et politique: un essai d'interprétation.* Paris: Quæ, 2003.
Peyton, Jonathan. "Imbricated Geographies of Conservation and Consumption in the Stikine Plateau." *Environment and History* 17 (2011): 555–81.
Pickstone, John V. "Les révolutions analytiques et les synthèses du modernisme." In *Histoire des sciences et des savoirs,* vol. 2, *Modernité et globalisation,* edited by Kapil Raj and Otto Sibum, 33–55. Paris: Seuil, 2015.
Piédalue, Gilles. "Les groupes financiers et la guerre du papier au Canada 1920–1930." *Revue d'histoire de l'Amérique française* 30, 2 (1976): 223–58.
Pisani, Donald J. "The Many Faces of Conservation: Natural Resources and the American State, 1900–1940." In *Taking Stock: American Government in the Twentieth Century,* edited by Morton Keller and R. Shep Melnick, 123–55. Cambridge: Cambridge University Press, 1999.
Pontbriand, Mathieu. *Lomer Gouin, entre libéralisme et nationalisme.* Sainte-Foy: Presses de l'Université Laval, 2016.
Pross, A.P. "Development of Professions in the Public Service: The Foresters in Ontario." *Canadian Public Administration* 10 (1968): 376–404.
Provencher, Jean. *La Station de recherche de Deschambault.* Quebec City: Ministère de la Culture et des Communications, 2006.
Rabinow, Paul. *Anthropos Today: Reflections on Modern Equipment.* Princeton, NJ: Princeton University Press, 2003.

Raffestin, Claude. *Pour une géographie du pouvoir*. Paris: LITEC, 1980.
–. "Space, Territory, and Territoriality." *Environment and Planning D: Society and Space* 30, 1 (2012): 121–41.
–. "Territorialité: concept ou paradigme de la géographie sociale?" *Geographica Helvetica* 2 (1986): 91–96.
Reigert, John. *American Sportsmen and the Origins of Conservation*. New York: Winchester Press, 1975.
Roby, Yves. *Les Québécois et les investissements américains, 1918–1929*. Montreal: Fides, 1976.
Rodgers, Andrew Denny, III. *Bernhard Eduard Fernow: A Story of North American Forestry*. New York: Hafner, 1968. First published in 1951 by Princeton University Press.
Rosanvallon, Pierre. *Le libéralisme économique: histoire de l'idée de marché*. Paris: Éditions du Seuil, 1989.
Rose, Nikolas, Pat O'Malley, and Mariana Valverde. "Governmentality." *Annual Review of Law and Social Sciences* 2 (2006): 83–104.
Rose-Redwood, Ruben. "Governmentality, Geography and the Geo-Coded World." *Progress in Human Geography* 30, 4 (2006): 469–86.
Rosenberg, Charles. "Toward an Ecology of Knowledge: On Discipline, Context, and History." In *The Organization of Knowledge in Modern America, 1869–1920*, edited by Alexandra Oleson and John Voss, 440–55. Baltimore: Johns Hopkins University Press, 1978.
Rossiter, David A. "Producing Provincial Space: Crown Forests, the State and Territorial Control in British Columbia." *Space and Polity* 12 (2008): 215–30.
Rousseau, Jacques. "Bataille de sextants autour du lac Mistassini." *L'Action universitaire* 14, 2 (1948): 99–116.
Rueschemeyer, Dietrich, and Theda Skocpol, eds. *State, Social Knowledge, and the Origins of Modern Social Policies*. Princeton, NJ: Princeton University Press, 1996.
Ryan, William F. *The Clergy and Economic Growth in Quebec (1896–1914)*. Quebec City: Presses de l'Université Laval, 1966.
Sack, Robert David. "Human Territoriality: A Theory." *Annals of the Association of American Geographers* 73, 1 (1983): 55–74.
–. *Human Territoriality: Its Theory and History*. Cambridge: Cambridge University Press, 1986.
Sahlins, Peter. *Boundaries: The Making of France and Spain in the Pyrenees*. Berkeley: University of California Press, 1989.
Saint-Hilaire, Marc. *Peuplement et dynamique migratoire au Saguenay 1840–1960*. Sainte-Foy: Presses de l'Université Laval, 1996.
Saint-Pierre, Jacques. *Les chercheurs de la mer: les débuts de la recherche en océanographie et en biologie des pêches du Saint-Laurent*. Sainte-Foy: IQRC, 1994.
Sandlos, John. *Hunters at the Margin: Native People and Wildlife Conservation in the Northwest Parks*. Vancouver: UBC Press, 2007.
Schulten, Susan. *Mapping the Nation: History and Cartography in Nineteenth-Century America*. Chicago: University of Chicago Press, 2012.
Scott, James C. *Seeing Like a State: How Certain Schemes to Improve the World Failed*. New Haven, CT: Yale University Press, 1997.
Séglard, Dominique. "Foucault et le problème du gouvernement." In *La raison d'État: politique et rationalité*, edited by Christian Lazzeri and Dominique Reynié, 117–40. Paris: Presses universitaires de France, 1992.
Séguin, Normand. *La Conquête du sol au XIXe siècle*. Montreal: Boréal Express, 1977.
Sivaramakrishnan, K. "A Limited Forest Conservancy in Southwest Bengal, 1864–1912." *Journal of Asian Studies* 56, 1 (1997): 75–112.
Snell, J.F. *Macdonald College of McGill University*. Montreal: McGill University Press, 1963.

Stafford, Robert A. "Geological Surveys, Mineral Discoveries, and British Expansion, 1835–1871." *Journal of Imperial and Commonwealth History* 12, 1 (1984): 5–32.

Strandsbjerg, Jeppe. "The Cartographic Production of Territorial Space: Mapping and State Formation in Early Modern Denmark." *Geopolitics* 13, 2 (2008): 335–58.

Sutherland Brown, A. "A Century of Service: A Brief History of British Columbia's Geological Surveys." *Geoscience Canada* 26 (1999): 31–37.

Thépot, André. "Les ingénieurs du corps des mines: évolution des fonctions des ingénieurs d'un corps d'État au XIXe siècle." *Culture technique* 12 (1984): 55–61.

Thibeault, Régis. *Industrie laitière et transformation agraire au Saguenay–Lac-Saint-Jean, 1870–1950.* Sainte-Foy: Presses de l'Université Laval, 2008.

Trépanier, Pierre. *Siméon Le Sage: un haut fonctionnaire québécois face aux défis de son temps (1867–1909).* Montreal: Bellarmin, 1979.

Vallières, Marc. "Les entreprises minières québécoises et la grande dépression des années 1930: le cas de la Noranda Mines." In *Private Enterprise during Economic Crises: Tactics and Strategies/L'entreprise privée en période de crise économique: tactiques et stratégies,* edited by Pierre Lanthier and Hubert Watelet, 133–47. New York, Ottawa, and Toronto: Legas, 1997.

–. "Le gouvernement du Québec et les milieux financiers de 1867 à 1920." *L'Actualité économique* 59, 3 (1983): 531–50.

–. *Des mines et des hommes: histoire de l'industrie minérale québécoise des origines au début des années 1980.* Quebec City: Ministère de l'Énergie et des Ressources du Québec, 1989.

Van Horssen, Jessica. *A Town Called Asbestos: Environmental Contamination, Health and Resilience in a Resource Community.* Vancouver: UBC Press, 2016.

Vandergeest, Peter, and Nancy Peluso. "Territorialization and State Power in Thailand." *Theory and Society* 24, 3 (1995): 385–426.

Vigod, Bernard. *Quebec before Duplessis: The Political Career of Louis-Alexandre Taschereau.* Montreal and Kingston: McGill-Queen's University Press, 1986.

Vincent, Odette et al., eds. *L'Abitibi-Témiscamingue.* Quebec City: IQRC, 1995.

Warren, Louis S. *The Hunter's Game: Poachers and Conservationists in Twentieth-Century America.* New Haven, CT: Yale University Press, 1997.

Whatmore, Sarah. *Hybrid Geographies: Natures, Cultures, Spaces.* London: Sage, 2001.

Winichakul, Thongchai. *Siam Mapped: A History of the Geo-body of a Nation.* Honolulu: University of Hawaii Press, 1994.

Wood, Denis, and John Fels. *The Natures of Maps: Cartographic Constructions of the Natural World.* Chicago: University of Chicago Press, 2009.

Zaslow, Morris. *The Opening of the Canadian North, 1870–1914.* Toronto: McClelland and Stewart, 1971.

–. *Reading the Rocks: The Story of the Geological Survey of Canada, 1842–1972.* Toronto: Macmillan, 1975.

Zeller, Suzanne. *Inventing Canada: Early Victorian Science and the Idea of a Transcontinental Nation.* Toronto: University of Toronto Press, 1987.

–. *Land of Promise, Promised Land: The Culture of Victorian Science in Canada.* Canadian Historical Association Historical Booklet 56. Ottawa: Canadian Historical Association, 1996.

Index

Note: "(f)" after a page number indicates a figure; "(t)" after a page number indicates a table.

abandoned land, 83, 84, 87–89
Abitibi region, 37, 56, 106, 115, 132; colonization of, 11, 60, 129; mining operations, 62–63, 65, 69, 70; township reserves, 91
Act for the Protection of Fisheries (1855), 102
Act to Facilitate the Formation of "Fish and Game Protection Clubs" (1885), 101
Adcock, Tina, xi–xii, xxii, 161*n*3
Agricultural Council, 125–26
Agricultural Instruction Act (1913), 24, 25, 128–30, 140, 143, 168*n*14
agricultural land, 73, 87–88, 134; separation of forestland and, 18, 36, 71, 72, 75, 79, 94
agricultural production: at Berthierville nursery, 87–88; commercial, 127–28; districts, 133–38, 135(f), 137(f), 140, 184*n*30; governmental interventions, 18–19; regionalization of, 34–35, 35(f), 132–33, 154; specialization of, 141–42; suitable crops for, 124; technoscientific interventions, 32(f), 32–35, 141, 154–55. *See also* dairy industry; fruit-growing industry; poultry production

agricultural societies, 3, 32, 125–26, 128, 141, 148, 154
agriculture schools and training, xvi–xvii, 23, 25, 28, 30, 154, 168*n*16; for aviculture, 148–49, 149–50, 152–54; educational speakers, 24, 32–33, 126–27; financing of, 33, 40
Agronomic Service, 19, 21, 23, 25, 30, 125, 129, 133, 139(f), 140, 143, 154
agronomists, 3, 19, 125, 154–55; data collection, 136–38; experimental activities, 34–35; geographic distribution, 130–32, 131(f); recruitment and growth, 25, 124, 128–29, 130(f); regional interventions, 132–34; reorganization of activities, 138, 139(f), 140; specialists, 140
Algonquin Park, 74
American Forestry Association, 73, 175*n*8
Anderson, Benedict, 163*n*36
angling. *See* fishing
apple production, 141, 142–44, 145(f), 147(f), 148
arboricultural stations, 34, 35, 142–46, 145(f)
asbestos extraction, 39, 48, 54, 56–57, 70; Thetford region, 27, 45, 58

200

autonomy, 42, 64, 70, 146, 155, 157, 167*n*77
aviculture, 149–51

Baldwin Mills, 117, 118, 120–21
Bancroft, Joseph Austin, 61, 64–65
Barlow, Alfred Ernest, 56, 58
Barnard, Édouard-André, 32, 126–28, 142
beavers, 115–16
Bédard, Avila, 26, 41, 80–82, 84–85, 87, 178*n*58
Belgo Pulp and Paper Company, 79, 80
Bell, Leslie V., 64
Bellisle, J.-Adolphe, 114
Berthierville nursery: agricultural activities, 87–88; education and training, 23, 81, 82(f), 84; establishment of, 35, 81; research activities, 37, 38(f); seedlings, 81, 85–86, 86(t), 92
Biays, Pierre, 89
Bignell, John, 52
Biological Service, 27, 28(f), 121(f), 122
Bois, Henri-Charles, 184*n*34
boundaries: electoral, 19, 134–36; Quebec territorial, 11
British Columbia, 14, 40, 41, 43, 166*n*59; mineral exploration, 46, 48
British North America Act, 44, 71, 99
Brunelle, Richard, 65
Bureau of Mines (ON), 15, 48, 62, 65
Bureau of Mines (QC), 15, 21, 48; activities in northwestern Quebec, 60–63; Division of Geology, 24, 27, 31, 39, 63–64, 69, 174*n*87; geological explorations, 50–51, 64–66, 65(f); hiring of personnel, 26–27, 70; laboratories, 37, 38(f), 39; map production, 17, 63, 64(f), 66–69, 68(f); mineral discoveries, 56–57; mineralogical analyses, 61–62

Canadian federal state, 3; administration, 3–5, 30, 39, 40, 41, 44, 79, 99, 117. *See also* Department of Agriculture (federal); Department of Marine and Fisheries (federal); Geological Survey of Canada
Canadian Forestry Association, 76, 79
Canadian Paper and Power Corporation (CPPC), 86–87, 92
Canadian Shield, 11, 44

Caron, Hector, 103–4, 111, 115–17
Caron, Joseph-Édouard, 128, 133, 136, 140
cartography. *See* maps/mapping
Castonguay, Stéphane, xii, xv, xvii–xxii
Catholic Church, 11
Chagnon, Stanislas-J., 184*n*34
Chambers, E.T.D., 117, 182*n*84
Chapleau, Joseph-Adolphe, 49, 169*n*32
chemistry, xvi, xvii
Chibougamau, Lake, 52, 54, 56–59, 61, 63, 69
Chibougamau Mining Commission, 58–59, 59(f), 63, 64
civil servants, 8, 13, 14, 15, 16, 72, 156; academically trained, 4, 157; Department of Agriculture, 19, 30; protection of fisheries, 102–3; total numbers, 20, 21; wildlife conservation, 97–98
Cobalt, 54, 56–57, 61
colonization, 40, 52, 87, 128, 158; of Abitibi region, 60, 129; forest resources and, 71, 73–76, 78, 89, 93–94; mineral resources and, 43–44, 50–51, 53; railway construction and, 11, 57–59
Colonization Commission, 74–75, 78
commercial fishing, 98, 102, 109, 180*n*3
Commission of Conservation, 30
Confederation, xv, 13, 14, 39, 43
conservation: fish, 98–99, 102–3; forest, 23, 72, 73–74, 78, 80, 175*n*8; movement, 72; of natural resources, 7, 26, 41; water, 76; wildlife, 29, 97–98, 101–2, 106–7, 110–12
Conservative governments: federal, 25; provincial, 13
cooperatives, 143, 147(f), 149–52
copper mining, 27, 45, 67, 70; Acton mine, 171*n*14; deposits, 47, 50, 60, 62
Cornell University, 79
corruption, 26, 79
Côte-Nord region, 37, 50, 60, 70, 95
Crown lands, 18, 22, 39, 41, 79, 112, 156; access to forest resources on, 74, 93–94; leases, 98, 101; reforestation of, 86, 89; revenues from, 4, 40, 43, 49, 72; survey, 75; watercourses, 99, 103, 122

dairy industry, 18, 138; cheese and butter factories, 23, 33, 127, 168*n*9; growth of,

125, 129, 154; regional specialization, 34, 136, 141; schools, 23
Daviault, Lionel, 37
Dawson, George Mercer, xvi, 45–46, 47(f)
Dawson, J.W., xvi
debt (public), 20, 22, 57
Denis, Bertrand-T., 64–65
Denis, Théophile-Constant, 31, 58, 59, 60, 62, 63, 66
Department of Agriculture (federal), Experimental Farms Branch, 30, 34, 79, 155, 163*n*23, 187*n*110
Department of Agriculture (QC), 18, 56; agronomic districts, 129–32, 131(f), 140; assistance to farmers, 132–33; Deschambault nursery, 35, 88, 143; educational enterprise, 32–33, 35, 124–27, 130, 154–55; funding for group activities, 29, 141; hiring of personnel, 23–25, 154; Horticultural Service, 143–46, 148; statistics section, 136, 185*n*43; support for regional production, 34, 125, 141–42. *See also* Agronomic Service; Horticultural Service; Plant Protection Service; Poultry Service
Department of Colonization, Mines, and Fisheries (QC), 36, 54, 57, 63, 75, 83, 88; hatchery service, 21, 39, 117. *See also* Bureau of Mines (QC); Fisheries and Game Service
Department of Crown Lands (QC), 4, 17, 20, 40; Survey Branch, 36; surveyors, 71. *See also* Crown lands
Department of Lands and Forests (QC), 21, 168*n*10; Berthierville nursery, 35, 80–81, 87–88; forest reserves, 36, 75–76, 78, 91, 92(f), 110; hiring of personnel, 23–24, 26, 79–80, 83; protection of wildlife resources, 111–12; use of scientific forestry, 93–94; Woods and Forests Branch, 26, 81–83. *See also* Forest Service
Department of Marine and Fisheries (federal), xv, 99, 116, 118
Department of the Interior (federal), 31, 47
Devlin, C.R., 58, 59
dispositifs, 5, 9, 10, 72, 141–42, 164*n*11
dissemination of information, 4–5, 158, 159; of agricultural knowledge, 19, 32–34, 125–29, 141, 154; of arboricultural techniques, 143; of forest protection, 36; to the mining industry, 48, 69
division of labour, 148, 149, 167*n*76
Dresser, John A., 31, 47, 64, 66
Duchesnay, H.J.J., 171*n*28
Dufresne, Alphonse-Olivier, 63–64
Dulieux, Émile, 51, 56, 60, 62
Dupré, Ruth, 160
Dupuis, Auguste, 142, 144

Eastern Townships, 69, 70, 75, 101, 144–46; hatcheries, 117, 118, 119–20; mineral deposits, 45, 47, 50, 60; poultry stations, 151–52
École d'agriculture de Sainte-Anne-de-la-Pocatière, xvi, 25, 128, 136
École de l'industrie laitière du Québec (Saint-Hyacinthe), 23, 33, 34, 127, 152
École des mines de Paris, 49
École normale de Québec, 149, 150, 152
École polytechnique de Montréal, 26, 29, 51, 60, 62, 66, 81; mineral analysis laboratory, 37, 39
economic crisis, 24, 91, 115, 134, 136, 141; back-to-the-land movement and, 104, 106; mining activities and, 62, 65
economic development, 12–14, 21
Elden, Stuart, 10
electoral ridings, 19, 140
entomology. *See* field laboratory; insects; Plant Protection Service
expeditions. *See* geological explorations
experimental farms, xvi, 30, 34, 79, 142, 149, 163*n*23
exploitation of natural resources, xii, xxi, 3–5, 106, 156–58, 160; of agricultural resources, 40; commissions of inquiry for, 41; of fish resources, 109; of forestlands, 15, 72, 73, 76, 93, 110; governmental interventions, xvii, 11, 12–15, 167*n*81; of mineral resources, 17, 44, 49, 60, 61, 69–70; revenues from, 20; surveying and, 36, 158, 159; of wildlife resources, 12, 15, 18, 95
explorers, 8, 36, 44, 45–46, 50. *See also* geological explorations

farmers: agricultural societies and, 125–26; assistance from the Department of Agriculture, 132–33; crop production and sales, 134; data supplied by, 136–38; education of, 40, 152–54, 168*n*16; proximity to agronomists, 128–32, 140–41; subsidies, 142, 150
farmers' club, 33, 126–28, 141
farming practices, 33, 124–25; for commercial production, 127–28
Fernow, Bernhard, 79
field laboratory, 35(f)
fieldworkers, xx, 26, 56, 123, 172*n*47
First World War, 24, 62, 97, 129
fish and game clubs, 12, 27; formation and land leases, 18, 98–101, 109–11; hatcheries and, 117; hiring of wardens, 102–4, 105(f), 106–7, 108(f); Laurentide National Park and, 112–13, 113(f); public revenues from, 97(f), 102, 123; reporting of information, 29–30, 95–96; targeting of species, 115–16; total number of, 97
fish hatcheries, 27, 96, 160; federal to provincial control, 39, 116–17; Saint-Faustin, 28(f), 121–22, 121(f); speckled trout and salmon, xv, 118–21
fish stocking, xv, 18, 27, 116, 123; researchers, 122; salmon fry, 118, 118(f), 120(f); watercourses, 39, 99, 121
Fisheries and Game Service, 20–21, 114, 180*n*5; recruitment of biologists, 18, 24; wardens, 103–4, 104(f), 105(f), 106, 107(f), 108(f). *See also* Biological Service; Hatchery Service
fishery protection, 98–99, 102–3, 107, 110
fishing: licences, 97, 98; practices, 106, 109, 114, 115; reserves, 115; rights, 18, 27, 98–100; tourist industry and, 96, 97, 117, 118–19
foreign capital, 13, 50, 51(f), 54
forest fires, 73, 76, 80, 84, 88–89
forest reserves: creation and mapping of, 17–18, 72–76, 77(f), 110; policy, 81; township, 78, 89, 90(f), 91–92, 92(f), 93(f). *See also* Parke Reserve
Forest Service, 17, 78; establishment and personnel, 23, 26, 82–83, 94; inventories,

36–37; reforestation activities, 72, 84–87, 88–89, 91–93
forestry schools, 41, 72; founding of Quebec's, 79–80; at Ontario universities, 79, 177*n*41; United States, 79, 177*n*38; Université Laval, 17, 23, 36, 81, 82(f); University of New Brunswick, 41, 79
forestry sector: colonization and lumbering conflicts, 17–18, 71, 73–75, 93–94; conservation and protection, 72, 76, 78, 93; development of Quebec's, 15; education and training, 79–83; inventories, xxi, 30, 36–37, 76, 163*n*1; provincial administrations, 30, 40–41, 79; recruitment of personnel, 23–24, 26; river destruction, 110, 181*n*51. *See also* forest reserves; reforestation
Fortier, Victor, 187*n*110
Fortin, Pierre-Étienne, 98
Foucault, Michel, xviii, xix, 10
fruit-growing industry, 19, 35, 88, 142–46, 145(f), 154

game clubs. *See* fish and game clubs
Gaspé, 37, 50, 66, 75, 101, 122, 132; arboricultural stations, 143, 146; geological explorations, 60, 69, 70; reserve, 111, 112, 115; salmon rivers, 95, 96, 99, 102
General Mining Act, 48–50, 62, 63, 69
geography, xviii–xix, 6, 15
geological explorations: in British Columbia and Ontario, 48; by Bureau of Mines geologists, 64–66; of Lakes Mistassini and Chibougamau, 52, 54, 56–59; for mineral deposits, 3, 46–47, 50–52; mineral maps of, 66–69, 67(f), 68(f); in northwestern Quebec, 27, 60–63, 69–70; student participants, 29, 66; territorial occupation and, 17; of unexplored territories, 45–46, 47(f); university professors under contract, 27, 29, 41, 51, 60–61, 65(f), 66
Geological Survey of Canada, 16, 26–27, 30, 50, 69; employment with, 29, 31; expeditions in northern Quebec, 52–54, 56; map production, 17, 66, 68; original mandate, xiv–xv, 43, 44–45; personnel

deployed in Quebec, 45–47; relocation to Ottawa, 49; report on mineral production, 48; unexplored territories and, 45–46; work in northwestern Quebec, 61, 62
geologists, 3, 26–27, 44, 48, 70; Geological Survey of Canada, 45, 46, 52, 56; Quebec Bureau of Mines, 17, 27, 58, 60, 64–66, 69
Godbout, Joseph-Adélard, 136, 138
gold mining, 27, 37, 61, 67, 70; deposits, 45, 47, 56, 60; inventories, 63; in Val-d'Or, 62
Gosse, Henry, xiv
Gouin, Lomer, 13, 26, 64, 75–76, 79–80
governmentality, xviii, 8–10
Grand Trunk Pacific Railway, 54, 55(f)
Grand-Mère, 85, 86
Gulf of St. Lawrence, 98–99

Hall, W.C.J., 74, 110
Harvie, Robert, 61
Hatchery Service, 27, 28(f), 39, 117–18, 120–21, 122–23
Horne, Edmund, 62
Horticultural Service, 143–46, 148
Hudson's Bay Company, 54
hunting, 27, 95, 102, 180*n*5; bans, 114, 115; practices, 106–7, 109; public, 115; regulations and rights, 100, 181*n*46; species targeted for, 115–16; territory, 100–1, 111–12, 116, 122
hydroelectricity production, 14, 75, 76, 167*n*81

ichthyological reports, 95, 117, 119, 122
Indigenous people, xi, xiv, 18, 98, 109
industrialization, 3, 13, 15, 28, 34, 93
infrastructure, 9, 110, 122; communication, 3; federal, 39, 40, 156; mining, 63; technoscientific, 28, 30. *See also* railway construction
Ingram, Darcy, 100, 109
insects, 3, 34, 37, 38(f)
inspection, 143, 155, 159; of cheese and butter factories, 23, 127, 168*n*9; forest, 73, 168*n*10; of lakes, 118

Institut agricole d'Oka, 25, 128, 149, 152, 168*n*16
institutional ecology, 15, 21, 167*n*76
interventionism, 14, 22, 30, 40, 157, 160
inventories, 7, 52, 63, 158, 159; forest, xxi, 30, 36–37, 76, 163*n*1

James Bay, 46, 52, 53, 54
Joncas, Louis-Zéphirin, 103, 110–12, 114
Jones, Islwyn Winwaloc, 64
Jones-Imhotep, Edward, xi–xii, xxii, 161*n*3
Journal d'agriculture illustré, 23, 32–33, 126–27, 187*n*110
jurisdiction, 6, 30, 99–100

Kinsey, Darin, 109
Kirkland Lake, 56, 61
knowledge, production of, 5, 25, 50, 156–58

La Vérendrye Reserve, 115
Laboratoire d'analyse officielle de la province de Québec, 31, 34
laboratories, 18, 31; field, 34, 35(f); forest research stations, 37, 38(f); mining, 37, 38(f), 39, 62, 63, 69, 70; in Quebec City, 16, 34
Lachute dunes, 84, 89
Lac-Saint-Jean (region), 11, 60, 63, 69, 89, 106, 119
Laflamme, J.C.K., 47, 51, 79–80, 81
laissez-faire economic system, 14
land agents, 7, 26
land reclamation, 78, 83, 87–89, 91
Langelier, Jean-Chrysotome, 82
Latour, Bruno, xiii, xv, 164*n*9
Laurentians (mountains), 136, 138
Laurentians (region), 60, 69, 84, 132; hatcheries, 28(f), 117, 118, 121(f)
Laurentide National Park: biology station, 122; forest reserve, 74, 89, 110; introduction of species to, 116; territory leased to private clubs, 100, 112–13, 113(f); total area, 112, 180*n*63; wildlife protection, 106, 112–13, 115
Laurentide Paper Company, 85; nursery, 87, 92, 93(f)
Le Lien, 134, 184*n*34

Le territoire, 12
leases, 13, 37, 87; fish and game territory, 18, 95, 96–101, 106; lakes and rivers, 103, 110, 114; Laurentide National Park, 112–13, 113(f); public reserves and, 110–11; timber limit holders and, 71, 72, 73, 76. *See also* timber limits
Legislative Assembly of Quebec, 58–59, 88, 100
Liberal governments (provincial), 13, 22, 93, 140, 166*n*62, 174*n*92
Liguori, Brother, 149
livestock raising, 127, 128, 138, 141, 154
Low, Arthur Peter, 52, 56, 57
Lower Canada, 11, 99, 102, 124, 125
Lower St. Lawrence, 91, 106, 144–46, 151, 186*n*84
lumbermen, 71, 73–75, 79, 81, 94
Lynch, William W., 73–74

Macdonald College, 25, 28, 128
Mackendie, C.D., 103
Made Modern (Jones-Imhotep/Adcock), xi–xii, xxii, 161*n*3
Magog, 117, 118, 119–20
Mailhiot, Adhémar, 60, 62
Manitou, Lake, 118, 121, 121(f)
maps/mapping, xix, 7, 17, 157, 158, 159; agronomic districts, 34, 131(f), 133–38, 135(f), 137(f), 154; apple production per county, 144–46, 145(f); distribution of fish and game wardens, 106, 108(f); distribution of salmon fry, 118, 119(f), 120(f), 121; forest and township reserves, 76, 77(f), 89, 90(f); geological exploration sites, 65(f); Laurentide National Park fish and game territory, 112–13, 113(f); mineral, 66–69, 67(f), 68(f); National Transcontinental and Grand Trunk Pacific railways, 54, 55(f); northwestern Quebec mining space, 61, 62, 63, 64(f); poultry stations by county, 151–52, 153(f); unexplored territory, 46, 47(f)
Marchand, Félix-Gabriel, 13, 22
markets, 124, 134, 154; urban, 138, 142, 148, 149

Martin, Paul-Louis, 144
Matane River, 181*n*51
Matapédia Valley, 104, 110, 132
McGill University, xvi, 29, 58, 66, 122, 129
Mercier, Honoré, 73
mineralogical analysis, 37, 38(f), 39, 61–62, 63, 171*n*28
mining industry, 15, 16–17, 30; colonization and, 43–44; development of northwestern Quebec's, 60–69; hiring of personnel, 26–27, 31; Ontario's, 54, 56–57, 60; ownership system, 48–49; promotion of Quebec's, 49–50, 51(f); railway construction and, 53–54, 57–60; surveying activities, 44–48; training and schools, 29, 36. *See also* geological explorations
Minville, Esdras, 16
Mistassini, Lake, 52, 54, 56–59, 63
modernization: agricultural, 12, 13, 127; of Canada, xi–xii, xvii; of Quebec state, xxi–xxii, 5, 42, 123, 158
Montreal, 53, 73, 138, 142, 151–52; agronomy districts, 130–32, 133; fruit producers, 144, 147
Murphy, Alexander, xviii–xix

national parks, 25, 72, 74, 110, 114, 115. *See also* Laurentide National Park; Parc de la Montagne-Tremblante
National Transcontinental Railway, 54, 55(f), 59, 60, 61, 129
natural history, xiii–xiv
natural resources industries, 12. *See also* exploitation of natural resources
naturalists, 4, 24, 26, 27, 50; Ottawa Field-Naturalists' Club, 45, 46
navigational technology, xiv
Nettle, Richard, 116
Noranda Mines, 62, 174*n*92
Northern Ontario Railway, 54
nurseries: Berthierville, 23, 35, 37, 38(f), 84, 85–86, 86(t), 87–88; Deschambault, 35, 88, 143; Laurentide (in Proulx), 86–87, 92, 93(f); locations for, 80–81; reasons for establishing, 80; for township reserves, 91–92, 92(f)

Obalski, Joseph, 26, 31, 61, 69; Lake Chibougamau exploration, 54, 56–57, 172*n*47; promotion of Quebec's mineral resources, 49–50, 51(f); retirement, 59

Ontario: forestry sector, 15, 41; hydroelectricity production, 14; mining industry, 54, 56–57, 60; prospectors, 54, 61, 62; public administration, 13–15, 39–40

orchards: commercial, 146, 148; demonstration, 34, 35, 143–46, 145(f), 150; spraying service, 147(f), 148

ore deposits, 3, 46, 50, 51, 63, 66; analysis of samples, 37, 39; in Appalachians region, 45; in northern Quebec, 53, 54, 56–57, 63; in Ontario, 15; ownership of, 49

O'Sullivan, Henry, 53, 54, 55(f)

Ottawa River, 73, 101, 110–11

Ottawa Valley, 45, 50, 72, 106, 151–52, 171*n*20

ouananiche. *See* salmon

Parc de la Montagne-Tremblante, 74, 110, 111

Parent, Simon-Napoléon, 13, 26, 75

Parke Reserve, 92(f)

Perrault, Joseph-Édouard, 64, 174*n*87, 174*n*92

Perron, Joseph-Léonide, 34, 141, 146, 148, 184*n*30; agricultural renewal program and map, 133–38, 140; cooperative plan, 151

phosphate, 45, 50, 171*n*20

Piché, Gustave-Clodimir: forestry school and training programs, 82–83, 177*n*57, 178*n*58; reforestation efforts, 87–89, 91, 93, 179*n*96; studies at Yale University, 26, 41, 79–80; tree nursery project, 80–81

Pickstone, John V., xiii–xiv

pisciculture, 18, 24, 96, 97, 116–17; stations, 27, 28(f), 118–22, 121(f)

Plant Protection Service, 34, 147(f). *See also* field laboratory

poaching, 103, 104, 106–7, 109, 111, 115

Pomological and Fruit Growing Society of the Province of Quebec, 142–43, 185*n*63

Pontiac County, 53, 56, 57, 59, 101; geological exploration, 61, 62–63, 69; public reserve, 110, 111

populations, 8–10, 18

poultry production, 19, 148–50; egg production, 149, 151–54, 152(f); growth by county, 151–52, 152(f), 153(f); stations, 141, 149–51, 153(f)

Poultry Service, 149–52

Prévost, Gustave, 121(f), 122

Prince, G.H., 41

private enterprise, 14, 61, 83, 85, 87

prospectors: Ontario, 54, 61, 62; Quebec, 36, 44, 45, 49, 50, 57

Proulx, J.-N., 103

Province of Canada, 3, 4, 15, 39, 116, 122, 125; debt, 20; geological portrait of, 43, 44; leases and, 96

public administration: Canadian, 3; federal institutions and, 30–31; growth of scientific personnel, 21–29, 31, 79; mining sector and, 44, 64; Ontario/Quebec comparison, 13–15, 39–40; private entities and, 95, 106; provincial differences, 41; specialization and defining spaces, xxi, 157, 158

public reserves, 96, 104, 110–13, 115, 122

pulp and paper industry, 17–18, 72, 83; demand, 93–94; forest resources and, 74–75, 78, 93(f); nurseries and reforestation projects, 80, 85–87, 94, 178*n*75; tree species and, 86

pulpwood forests, 18, 21, 74; production of, 78, 85–86, 89, 93(f), 93–94

Puyjalon, Henri de, 95

Quebec and Lake St. John Railway, 53, 101, 113(f)

Quebec Boundary Extension Act (1898), 52

Quebec City, 22, 24, 53, 147; laboratories in, 16, 34, 39, 69; poultry stations, 151–52

Queen's University, xvii, 29, 79, 177*n*41

Raffestin, Claude, xviii–xix, 6–7

railway construction, 3, 13, 14, 22, 46, 70; to Chibougamau region, 58–59; colonization and, 11, 53, 57; depletion of forests and, 71; Lake St. John and James Bay proposal, 53–54; National Transcontinental Railway, 54, 55(f), 59, 60, 61, 129

rainbow trout. *See* trout

reforestation, 23, 40; abandoned land for, 88–89; Forest Service projects, 72, 84–87; township reserves, 18, 78, 89, 90(f), 91–92, 92(f), 93(f); training programs, 17, 83, 84
representations, 6–8, 9, 69, 157
reserves. *See* forest reserves; public reserves
Richard, Louis-Arthur, 115
Richelieu Valley, 132, 146
Rioux, Albert, 140
road construction, 14, 100, 133
Rosenberg, Charles, 167*n*76
Roy, Louis-Philippe, 184*n*34
Rumilly, Robert, 169*n*32

Sack, Robert, xviii–xix, 6–7
Saint-Faustin, 28(f), 121(f), 121–22
Saint-Félicien, 118, 120
Saint-Maurice Valley, 53, 60, 72, 85, 95, 101
salmon: Atlantic fry, 117–18; decrease in stocking activities, 120–21; distribution of freshwater salmon *(ouananiche)*, 118–19, 119(f), 120(f), 121; regeneration of, xv, 116; rivers, 95, 96, 98–99, 102; timber industry and, 110
Savoie, François Narcisse, 133, 140
science and technology, xiii–xv, xxi–xxii; fieldworkers, xx; institutionalization of, xv–xvii, 4–5. *See also* technoscientific activities
scientific forestry, 17, 72, 78, 79–82, 87, 94
scientific services, 4–5, 12, 16, 40, 42; diversification of activities, 31–37, 39–42; government departments, 20–21; growth of personnel, 21–29, 42, 157; specialization of, 158. *See also* technoscientific activities
Scott, James, 9
seedlings: Berthierville tree nursery, 81, 85–86, 86(t), 92; coniferous, 37; distribution of, 34, 88; fruit tree, 34–35; for pulp and paper companies, 80; for township reserves, 91–92, 92(f)
settlers, xii, 40, 71, 106; colonization regions, 11, 53; forest reserves and, 73–76, 78; hunting rights, 100; township reserves and, 17–18, 89, 91
silver deposits, 54, 56, 63

small-mouth bass, 119–20
Société de géographie de Québec, 52, 53
Société d'industrie laitière de la province de Québec, 33, 127, 141
soil: agricultural potential of, 11, 71, 134; analysis, 18, 34; classification, 36, 75, 83, 93–94, 158; fertility, 84, 87–88, 124; improvement, 134; subsoil, 48–50, 61, 62, 69–70
sovereignty, 6, 7, 10
space: forest, 73, 78, 94; governmentality and, 10, 165*n*43; mining, 44, 51, 52, 56–57, 63, 64(f), 69; public administration and defining, xxi, 157, 158; state and, xix, 6–8; territoriality and, xviii–xix
speckled trout. *See* trout
sportfishing, 27, 39, 102, 116–18, 180*n*3; development of, 109–10; regulations, 98
sportsmen. *See* fish and game clubs
St. John, Lake (Lac-Saint-Jean), 89, 101, 106, 132; fish stocking, 118, 118(f); geological explorations, 52–53, 57; railway line, 58, 110
St. Lawrence Lowlands, 44, 136
St. Lawrence River, 47, 60, 74, 80, 100, 121
state formation, xviii, 5; governmentality, 8–10
state power, 6, 8–9, 42, 156, 157
surveillance, 96, 98, 122, 158; of fish and wildlife territory, 18, 27, 100, 102–4, 106, 108(f); of forestry activities, 81, 88; park access and, 114
surveys, 7, 16, 32, 158, 159, 163*n*1; Lake Mistassini, 52; land, 17, 36, 71, 72, 75; ore deposits, 50, 53; railway construction, 53–54, 55(f). *See also* Geological Survey of Canada

Taché, Joseph-Charles, 124, 125, 154
Taschereau, Alexandre, 13, 64, 133, 174*n*92
Taylor, Bertram William, 24, 27, 117–20, 118(f), 122
technoscience, concept and definition, xiii–xv, 164*n*9. *See also* science and technology
technoscientific activities, 4–5, 10, 18, 156; agents, xx–xxi, xxii, 158; agricultural resources and, 32(f), 32–35, 141, 154–55;

expenditures for, 31(t), 159–60; exploitation of resources and, 15, 16; external service personnel, 24(t), 24–27, 29; forestry, mining, and wildlife resources and, 33(f), 36–37, 38(f), 39–40; internal service personnel, 22(t), 22–24; provincial differences, 40–41; role of, 8
Témiscamingue, Lake, 54, 56–57, 61
territoriality, xviii–xix, 6, 11, 16, 167n81
territory: agricultural, 125, 132, 134–36, 154; colonization and, 53, 57; fish and game, 18, 96–102, 110, 112, 116, 122–23; governmentality and, 8–10; mapping and surveying, xix, 7, 36, 157, 158; mineralogical resources and, 17, 61; occupation of, xxi, 5, 7, 8, 12, 14, 60, 91; of the province of Quebec, 11–12; state power and, 6, 9, 156, 157; surveillance of, 102–4, 106; unexplored, 45–46, 47(f), 50
Thetford region, 27, 37, 45, 56–57, 58, 66
timber limits, 7–8, 36, 71, 72, 83, 156–57; delimiting, 17; fires in, 88–89; forest reserves and, 76, 91, 110; holders and settler disputes, 18, 72–75; reforestation of, 80, 84–86, 88, 94
Toumey, James William, 177n45
Tourchot, Léon Anatole, 31
tourism industry, 96, 97, 109, 114, 117, 119
tree nurseries. *See* nurseries
tree species, 18, 72, 75, 81, 85–86, 86(t), 92
Triton Club, 100, 101, 112, 180n23
Trois-Rivières, 36, 117, 138, 152
trout, 116–22

Union expérimentale des agriculteurs de Québec, 148–50
Union nationale government, 16, 29, 140

Université de Montréal, 122
Université Laval, 66; agriculture school, 25, 128; École de foresterie (forestry school), 17, 23, 26, 36, 81, 82(f), 83, 84, 88; Montreal branch, 25
universities, xvi–xvii, 4, 28–29, 41; granting of degrees, 25, 35, 64, 169n38. *See also names of institutions*
University of Toronto, xvi, 41, 79, 177n41

veterinary schools, 25
Vladykov, Vadim D., 122

wardens, fish and game: distribution of, 106, 107(f), 108(f); fines collected by, 105(f); hired by the Fisheries and Game Service, 104, 104(f), 106; forest, 82; private clubs vs Fisheries and Game Service, 105(f), 106; role of, 18, 95–96, 99–100; salaries, 103–4; surveillance by, 27, 111–12
watercourses, 18, 54, 66, 98, 117; on Crown lands, 99, 103, 122; fish stocking in, 39, 118–21; preservation of, 112; timber industry and, 110
Weber, Max, 162n7
white spruce, 85–87, 92
wildlife resources: conservation or protection of, 29, 74, 97–98, 100–2, 106, 110–13; depletion of, 102; exploitation of, 12, 15, 18, 95; reports and surveys on, 95–96
Wilmot, Samuel, xv, 116
Wilson, Ellwood, 85–87
Wynne-Edwards, V.C., 122

Yale University, 26, 41, 79–80, 177nn44–45

Zeller, Suzanne, xvii

Claire Elizabeth Campbell, *Shaped by the West Wind: Nature and History in Georgian Bay*
Tina Loo, *States of Nature: Conserving Canada's Wildlife in the Twentieth Century*
Jamie Benidickson, *The Culture of Flushing: A Social and Legal History of Sewage*
William J. Turkel, *The Archive of Place: Unearthing the Pasts of the Chilcotin Plateau*
John Sandlos, *Hunters at the Margin: Native People and Wildlife Conservation in the Northwest Territories*
James Murton, *Creating a Modern Countryside: Liberalism and Land Resettlement in British Columbia*
Greg Gillespie, *Hunting for Empire: Narratives of Sport in Rupert's Land, 1840–70*
Stephen J. Pyne, *Awful Splendour: A Fire History of Canada*
Hans M. Carlson, *Home Is the Hunter: The James Bay Cree and Their Land*
Liza Piper, *The Industrial Transformation of Subarctic Canada*
Sharon Wall, *The Nurture of Nature: Childhood, Antimodernism, and Ontario Summer Camps, 1920–55*
Joy Parr, *Sensing Changes: Technologies, Environments, and the Everyday, 1953–2003*
Jamie Linton, *What Is Water? The History of a Modern Abstraction*
Dean Bavington, *Managed Annihilation: An Unnatural History of the Newfoundland Cod Collapse*
Shannon Stunden Bower, *Wet Prairie: People, Land, and Water in Agricultural Manitoba*
J. Keri Cronin, *Manufacturing National Park Nature: Photography, Ecology, and the Wilderness Industry of Jasper*
Jocelyn Thorpe, *Temagami's Tangled Wild: Race, Gender, and the Making of Canadian Nature*
Darcy Ingram, *Wildlife, Conservation, and Conflict in Quebec, 1840–1914*
Caroline Desbiens, *Power from the North: Territory, Identity, and the Culture of Hydroelectricity in Quebec*
Daniel Macfarlane, *Negotiating a River: Canada, the US, and the Creation of the St. Lawrence Seaway*

Justin Page, *Tracking the Great Bear: How Environmentalists Recreated British Columbia's Coastal Rainforest*

Ryan O'Connor, *The First Green Wave: Pollution Probe and the Origins of Environmental Activism in Ontario*

John Thistle, *Resettling the Range: Animals, Ecologies, and Human Communities in British Columbia*

Jessica van Horssen, *A Town Called Asbestos: Environmental Contamination, Health, and Resilience in a Resource Community*

Nancy B. Bouchier and Ken Cruikshank, *The People and the Bay: A Social and Environmental History of Hamilton Harbour*

Carly A. Dokis, *Where the Rivers Meet: Pipelines, Participatory Resource Management, and Aboriginal-State Relations in the Northwest Territories*

Jonathan Peyton, *Unbuilt Environments: Tracing Postwar Development in Northwest British Columbia*

Mark R. Leeming, *In Defence of Home Places: Environmental Activism in Nova Scotia*

Jim Clifford, *West Ham and the River Lea: A Social and Environmental History of London's Industrialized Marshland, 1839–1914*

Michèle Dagenais, *Montreal, City of Water: An Environmental History*

David Calverley, *Who Controls the Hunt? First Nations, Treaty Rights, and Wildlife Conservation in Ontario, 1783–1939*

Jamie Benidickson, *Levelling the Lake: Transboundary Resource Management in the Lake of the Woods Watershed*

Daniel Macfarlane, *Fixing Niagara Falls: Environment, Energy, and Engineers at the World's Most Famous Waterfall*

Angela V. Carter, *Fossilized: Environmental Policy in Canada's Petro-Provinces*

Printed and bound in Canada by Friesens
Set in Garamond by Artegraphica Design Co. Ltd.
Copy editor: Francis Chow
Proofreader: Kristy Lynn Hankewitz
Indexer: Celia Braves
Cartographer: Émilie Lapierre Pintal, CIÉQ *(unless otherwise noted)*
Cover designer: Lara Minja
Cover image: Rain gauge at Berthierville nursery, 1942 | Photo by T. Deslauriers | BANQ-Québec, Fonds ministère de la Culture et des Communications, Office du film du Québec, Documents iconographiques, E6, S7, SS1, P6762